WHAT
INVERTEBRATE?
A BUYER'S GUIDE
FOR MARINE AQUARIUMS

Tristan Lougher

I N T   H I N G

© 2007 Interpet Publishing,
Vincent Lane, Dorking, Surrey
RH4 3YX, England.

ISBN: 978 1 84286 179 0

Credits

Created and compiled:
Ideas into Print, Claydon,
Suffolk IP6 0AB, England.

Design and prepress:
Philip Clucas MCSD

Production management:
Consortium, Poslingford, Suffolk
CO10 8RA, England.

Print production: Sino Publishing
House Ltd., Hong Kong.

Printed and bound in China.

The star rating ★★★★★

We include a guide to the prices that aquarists might expect
to pay for individuals of each species. The range of prices for
each group of invertebrates is given on the opening page of
the section. Sometimes, the star rating will cover more than
one price category. Invertebrates, particularly corals, can be
available as very small or very large colonies. The larger the
invertebrate, the greater the cost of importing or growing it.
Some invertebrates, such as pulse corals, use up oxygen at a
fast rate when bagged for shipping and thus need accordingly
large bags to allow for this. The larger the invertebrate, or the
bag that holds it, the greater the cost of importing it. The final
price of an invertebrate may also reflect where the dealer has
purchased it. Directly imported invertebrates may command a
lower price than those sourced through wholesalers. Similarly,
captive-bred or propagated individuals may actually cost
less than their wild-collected counterparts, as their asexual
reproduction is a by-product of a successful aquarium
and aquarists are often only too happy to rid themselves
of rapidly spreading stock. It is unwise to select a coral or
other invertebrate simply based on its price; first establish its
condition and suitability for the aquarium.

Author

Tristan Lougher has been keeping aquariums for over 25 years,
and his interest in all things animal led to a degree in zoology
in 1992. Since then he has worked as a professional aquarist,
mainly with Cheshire WaterLife Ltd, where he specialises
in marine fish and invertebrates. He has written for several
publications, including Practical Fishkeeping, Today's Fishkeeper,
Tropical World and Tropical Fish.

   He currently contributes to Marine World, where his articles
include the very popular 'Friend or Foe' column that identifies
invertebrates that are accidentally introduced into reef aquariums,
and a feature called 'Spineless Wonders', in which he describes
rare or expensive species of invertebrates and their husbandry.
Tristan is interested in all aspects of the aquarium hobby and
zoology, including palaeontology. He is also an enthusiastic
SCUBA diver and tries to get away to the Red Sea at least
once a year.

Introduction

In common with marine fish, invertebrates for the home saltwater aquarium are represented by the beautiful, the dangerous, the easy-to-care-for and the problematic. This book aims to provide information concerning the basic demands of marine invertebrates before you are tempted to make a purchase. Modern aquariums housing invertebrates are almost exclusively based around live rock and this book assumes that this is so.

Many invertebrates harbour photosynthetic symbionts (zooxanthellae) that supplement their dietary needs, whereas others may demand regular feeding. The main sources of nutrition for each species are discussed, together with the best feeding methods. Where invertebrates are to be kept with fish, we look at their compatibility and whether they have any potential to do harm to each other. There is also a general section for additional information that is difficult to pigeon-hole and yet should help you avoid potential pitfalls.

It is also worth appreciating how large a specimen can become. For many species their maximum growth potential will be larger than the aquarium itself, in which case it is important to know whether you can take cuttings to reduce their final size or to provide others with specimens.

As well as a guide to the price you might expect to pay, and the point of origin of specimens, we also try to identify species that are similar in appearance to the animal in focus. This can give you alternative ideas for stocking or avoid the acquisition of a troublesome specimen through misidentification.

Finally, in common with their vertebrate relatives, many of the spineless additions to our aquariums confound even the best advice regarding their husbandry. Although the information in this book is based on years of personal experience, plus that of friends and a large number of reef aquarists, there are exceptions to every rule. In writing this book I am forced to deal in generalisations and readers should be aware that rogue individuals can and do occur.

Tristan Lougher

Contents

In this listing, the common name is followed by its scientific name. In sections featuring more than one group of ivertebrates, the species are presented in A-Z order of scientific name within each group.

▶ *Hardy and beautiful but packing a potent sting: Tube anemones from the genus* Cerianthus.

▲ Faviid stony corals often contain fluorescent pigments that aquarists find highly attractive.

62 – 81

Stony corals

Contents

82 – 91

Gorgonians

▼ *Recent advances in invertebrate diets mean that beautiful species, such as this blue polyp sea fan (Acalyclgorgia sp.) are no longer impossible to keep in the home aquarium.*

92 – 127

Crustaceans

▲ Fire shrimps (Lysmata debelius) *are hardy favourites of marine aquarists.*

128 – 149

Molluscs

Contents

▼ *Polychaete fanworms, known as featherdusters, are inexpensive animals that require feeding if they are to thrive.*

184 – 193

Miscellaneous

194 – 205

Accidental arrivals

206

A-Z species list

208

Picture credits and acknowledgements

How to use this book

▶ This book is designed as a buyer's guide to tropical marine invertebrates and, as such, assumes that readers have established an aquarium that contains all the filtration components required to ensure good water quality and to provide stable conditions in the long term. Any special considerations for a particular parameter, are discussed under the appropriate heading. Every organism has a species, or scientific, name. It is very useful to familiarise yourself with these, as one animal can have several common names. Some species with identical common names have very different aquarium requirements, so distinguishing between them is vital. The common names given here are those most often encountered in retail outlets. On these pages we describe how the book is organised and the information you will find under each heading.

Price guide

★	The price
★★	ranges are
★★★	calibrated for
★★★★	each section
★★★★★	of the book.

PROFILE

Each species entry begins with a brief profile, describing the essential character of the creature, any points of particular interest and its value and role in the aquarium.

WHAT size?

For most animals it is sufficient to describe how large an individual will become in terms of its length or diameter, and this has been done for relevant species. However, as many marine invertebrates are colonial in nature and can reproduce asexually, we have also given an indication of how large a particular colony may grow. For example, an individual *Discosoma* spp. mushroom polyp may reach 15cm in diameter, but through asexual reproduction a colony of many individuals may arise, each measuring 15cm but resulting in a total spread of many metres!

WHAT does it eat?

All aquarium animals require sustenance and this section describes the types of food that a particular animal will consume. For those species that contain photosynthetic symbionts (zooxanthellae), feeding may not be strictly necessary. However, almost every tropical marine aquarist aims to recreate nature in their living room, and offering animals supplementary foodstuffs reflects their natural state more closely. In the wild, they supplement the products of photosynthesis with particles captured from the water.

WHERE is it from?
The tropical ocean or sea in which the animal is naturally encountered is listed, together with information on the major collection areas, where known.

WHAT does it cost?
★★★☆☆
The price of a specimen reflects its size, origin, where the retailer purchased it, whether it is cultivated or wild-collected and the difficulties associated with its transport. In the case of some species, such as certain corals, which are available both as very small and very large specimens, the stated price range is quite broad. Price can also reflect the

colours and rarity of a specimen. For example, brown and green bubble anemones *(Entacmaea quadricolor)* are more common than red specimens of the same species. Consequently, the latter command a much higher price premium.

HOW many in one tank?
It is important to know whether individuals of the same species are compatible or not. Some will thrive when kept in numbers and, in some cases, it is possible to observe reproduction or courtship when more than one individual is present. In other species, individuals may be highly territorial and it is vital to keep only a single specimen.

HOW compatible with other invertebrates?
This section tells you what to look out for when siting a

SIMILAR SPECIES
So many invertebrates could have been included in this book that it is only possible to discuss a representative sample of those most commonly encountered. In these panels, the emphasis is on species not included elsewhere. Where possible, we describe the characteristics to enable you to identify between two or more species. Where it is particularly important that two species are not confused, for example when they share a common name, they are discussed under this heading.

specimen and describes any potential conflicts that may occur. In much the same way as plants compete for light in a garden, so corals compete for this most valuable resource in the aquarium. Some grow quickly and put their neighbours in the shade, whereas others use their powerful stinging cells to attack competitors directly. Some even resort to chemical warfare to oust their rivals. And, of course, certain invertebrates actively predate other ones.

WHAT water flow rate?
Under this heading we highlight the requirements of animals

◄ *Invertebrates in a mature reef aquarium create a natural appearance.*

that are influenced by water flow. In their natural environment, corals will grow from larvae or through asexual means into a shape that is governed by the conditions that are local to it. Apart from light, one of the main influences on a coral will be the water currents that act upon it. This is particularly true of stony corals that grow by secreting a skeleton that, once in position, is unlikely to move voluntarily. In areas of relatively low water flow, such corals can often put their energies into rapid growth and consequently their skeleton is less dense than one found in turbulent water. The latter may have densely packed branches and a very robust appearance.

Water flow around corals can influence not only their shape, but also has an important role to play in facilitating gas exchange and removing waste products from the coral. If a fast-growing yet brittle coral is placed in an area where the flow is too strong, it could be broken or severely damaged. Conversely, a more robust coral, originating from a high flow area, may not have sufficient flow around it to enable it to function properly and it, too, will not thrive. In the case of most stony corals it is true to say that if they are not positioned correctly with regard to their water current requirements, specimens are likely to die.

HOW much light?

Lighting provides the means by which we can observe the animals in our aquarium.

▲ *Lighting is a vital part of a system in which invertebrates containing photosynthetic symbionts are kept.*

▲ *Learn the aquarium needs of corals so you can position them according to their light and water current demands.*

Lighting (and water currents) are not absolutely essential for every species of invertebrate described in this book, but for the corals in particular they are of paramount importance. Many marine animals contain symbiotic algae called zooxanthellae that require lighting of the correct intensity and wavelengths in order to thrive. Many corals also produce special proteins that are able to convert harmful radiation, typically from the ultraviolet range of the electromagnetic spectrum, into photosynthetically available radiation (known as PAR). It is the

combination of these protective proteins and the numbers of zooxanthellae present that endow corals with their attractive appearance.

Corals are able to manipulate the numbers of zooxanthellae they contain according to their environmental conditions, typically by rationing the nutrients they supply to these symbionts. In some cases, particularly when the aquarium environment is unsuitable, they are unable to control the supply of nutrients and the zooxanthellae can reproduce apace. Often this results in the 'browning' of hitherto colourful corals. Light is, therefore, extremely important to most reef aquarium residents in terms of its intensity and composition.

In the book we mention three types of lighting. T8 lamps are 25mm-diameter fluorescent tubes, available from a number of manufacturers in a range of different colour temperatures (measured in degrees Kelvin). A 5000K light will have a warmer, pinkish glow compared to the crisp white output of a 13000K lamp. 20000K lamps appear blue to the observer and can be used to simulate deeper water environments. Such lamps are also useful for highlighting fluorescent pigments and can actually stimulate the coral to produce more of these attractive proteins. T8 lamps need to be used in numbers if they are to be of use to corals requiring strong illumination.

T5 fluorescent lamps are

16mm in diameter and have roughly twice the power output of a T8 lamp. They are also available in compact form. Aquarists have created beautiful aquariums – homes to animals that demand very strong illumination – using T5 lamps, but they have employed multiple lamps of high wattages to achieve these results.

Both forms of fluorescent lighting tend to result in an aquarium with quite a flat appearance that some aquarists consider 'unnatural', although this is a matter of preference.

Metal-halide remains the lighting of choice for many reef aquarists. Its high-output lamps are available in sizes up to 1000 watts each, with colour temperatures suitable for the requirements of the most demanding hobbyists. They most closely simulate the effect of natural sunlight as it penetrates water, producing areas of more concentrated light caused by diffraction at the surface.

Light intensity diminishes rapidly as the rays descend through water. Thus animals requiring more light might need to be placed towards the top of the aquarium, unless the wattage of the light unit(s) is sufficient. Where possible and relevant, we indicate the intensity of lighting that should fall upon a particular coral.

WILL fish pose a threat?

As this book is intended for aquarists maintaining mixed fish and invertebrate aquariums, we assume that non-reef-compatible

▲ *Common clownfish have adopted a euphyllia coral as a surrogate anemone.*

fish species are not housed. However, fish stocks that will peacefully coexist with many species of invertebrate may have a tendency to nibble at particular species and, where possible, we have attempted to name and shame the perpetrators. For example, many aquarists house true angelfish, such as the regal *(Pygoplites diacanthus)* in a reef aquarium, as they will generally ignore most commonly kept corals. However, they do have a tendency to nibble at the tube feet of starfish and, occasionally, at some large-polyp stony corals. Thus, they are generally safe to house in an aquarium where these animals are absent. It is worthwhile pointing out that rogues in any fish species can and do occur and this information is meant only as a guideline.

WILL it threaten fish?

As the question suggests, this is where we discuss the presence and nature of any threats to fish that may be inflicted by the invertebrate being described.

WHAT to watch out for?

Under this heading you will find any important points that you should consider before buying a particular invertebrate. It may include additional information not covered by the specific title headings or it may describe some features that will help you to distinguish between healthy and ailing specimens.

WILL it reproduce in an aquarium?

Where known, we have supplied information concerning the potential for asexual and sexual reproduction of each species.

Simple but fascinating

Sponges belong to the phylum Porifera and represent the first evolutionary steps in single-celled organisms organising themselves into multicellular animals. In this respect they have been highly successful for millions of years and of over 8000 species currently described, the vast majority are marine in origin. Primarily adapted to filter organic particles and plankton from the water column, some species have also incorporated photosynthetic algae into their tissues. Since they rely on an unobstructed surface area, problems can arise in aquariums where high nutrient levels fuel the growth of algae that may foul the animals. Large amounts of detritus may also compromise the ability of these animals to feed. However, given the right types of food on a regular basis, many sponge species can be kept successfully in a well-maintained system.

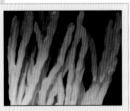

Price guide

★	£12-18
★★	£12-25
★★★	£20-35

PROFILE

The spiny sponge is so-named because of its jagged appearance. The same name is given to a number of species of sponge from the Indo-Pacific. In fact, many different species of sponge have this overall appearance, their shape being influenced by currents and grazing pressure.

WHAT size?
Usually offered for sale at about 10-20cm, but can achieve double this or more.

WHAT does it eat?
Requires regular feeding with fine particulates such as phytoplankton.

WHERE is it from?
Tropical Indo-Pacific.

WHAT does it cost?
★★☆☆☆
Price depends on the size of the animal.

Spiny sponge

SIMILAR SPECIES

Yellow forms are commonly offered for sale. There are many species with a similar growth form, including *Dendrilla* and *Stylissa* spp., but their correct identification to species level is almost impossible without microscopic analysis of their spicules.

WILL fish pose a threat?

May be nibbled by fish, including true angelfish, for whom sponges form part of their natural diet. However, this behaviour is rare due to the animal's chemical defences.

WILL it threaten fish?

Should not affect any fish species.

WHAT to watch out for?

Avoid individuals with clear areas around their ridges and peaks at the extremities of the colony. These signify dying flesh. Always ensure that an animal has been well settled into the dealer's aquarium before buying it. Do not expose the sponge to air, as bubbles can form in the tissue when it is immersed. This will result in the death of the surrounding area.

WILL it reproduce in an aquarium?

Unlikely.

⬆ *The branched growth of spiny sponges can trap detritus and provide a large surface area for algal growth. Prudent use of water currents and low nutrient levels will help to reduce problems.*

HOW compatible with other invertebrates?

This animal secretes noxious chemicals that can affect the growth of its neighbours.

WHAT water flow rate?

Enjoys strong water currents that help to keep its surface free from detritus and sediment.

HOW much light?

Does not enjoy becoming smothered with encrusting algae. The risk can be reduced by placing specimens in shaded areas of the aquarium and controlling algal nutrients, such as nitrate and phosphate.

Clathria rugosa

Orange fan sponge

PROFILE

This species is readily available and often bought for its beautiful coloration. However, its fairly straightforward requirements are seldom met and it is frequently abused. It is intolerant of algae that can overgrow it, especially where levels of phosphate and nitrate are high in the water.

WHAT size?
In its natural environment, the fan sponge can reach one metre or more in diameter. It is offered for sale at 15cm or so.

WHAT does it eat?
Consumes very fine particulate material. Bacteria probably form the bulk of the diet of wild specimens, but phytoplankton is best in the aquarium. Add it daily. Many aquarists drip phytoplankton into the aquarium overnight.

WHERE is it from?
Tropical Indo-Pacific. Similar species occur circumtropically.

WHAT does it cost?
★☆☆☆☆
Price depends on the size of the animal.

▲ *This inexpensive yet colourful sponge should have solid colours with no clear or pale areas. It is often purchased for its colour alone, with no regard given to its specialised needs.*

HOW compatible with other invertebrates?
This animal secretes noxious chemicals that can affect the growth of its neighbours.

WHAT water flow rate?
Strong currents are absolutely vital for this species and its close relatives. Some authorities state that unidirectional, constant flow is essential, but it should be high-flow, low-velocity.

HOW much light?
Less tolerant of direct illumination than other species in this section.

The animal itself has no real problems, but given that algae can easily smother it, place it in shadier areas of the aquarium.

WILL fish pose a threat?
May be nibbled by fish, including true angelfish, for whom sponges form part of their natural diet. However, this behaviour is rare due to the animal's chemical defences.

WILL it threaten fish?
This sponge should not harm fish.

WHAT to watch out for?
These sponges deteriorate slowly. Allow them time to settle down in the dealer's aquarium before buying them. They should be a uniform orange colour, with no pale patches. Take a close look at the outer margin of the fan and ignore specimens with signs of clear tissue in this area, no matter how insignificant it may appear. Do not expose the sponge to air, as bubbles can form in the tissue when it is immersed. This will result in the death of the surrounding area.

WILL it reproduce in an aquarium?
Unlikely.

SIMILAR SPECIES
There are several species of fan, or paddle, sponge. Some will be from the genus *Clathria*, others may be from the genera *Axinella* or *Pseudaxinella*.

Haliclona sp.

Blue tubular sponge

PROFILE

A beautiful animal, prized for its vivid blue colour, but widely misunderstood by aquarists. Far from being intolerant of bright light, it thrives in such conditions and is often found on rocks that are home to stony corals that require strong illumination.

WHAT size?
Can grow very large. Specimens offered for sale will range between 10 and 20cm in diameter, but there is no theoretical upper limit for growth in the aquarium, providing they have space and substrate to exploit.

WHAT does it eat?
Feed daily on fine particulate material, including phytoplankton and microplankton.

WHERE is it from?
The genus is circumtropical in distribution, but the tubular blue form is mainly collected from the Indo-Pacific region.

WHAT does it cost?
★★★☆☆
Price depends on the size of the animal.

▶ *Blue sponge is often associated with photosynthetic hard corals, proving that it can thrive under moderate to strong illumination.*

HOW compatible with other invertebrates?
Fairly aggressive. Has the potential to overgrow other sessile animals in the aquarium or irritate them with toxic secretions.

WHAT water flow rate?
Strong currents are essential for the long term well-being of this sponge.

HOW much light?
Does not require any light, but is able to withstand direct illumination.

WILL fish pose a threat?
May be nibbled by fish, including true angelfish, for whom sponges form part of their natural diet. However, this behaviour is rare due to the animal's chemical defences.

WILL it threaten fish?
Should not harm fish.

SIMILAR SPECIES
The purple sponge *Haliclona molitba* is similar in form but readily distinguished by its characteristic coloration (see page 18).

WHAT to watch out for?
Avoid specimens that appear pale or have discoloured patches in the vivid blue pigment. Patches of brown tissue indicate that these areas are dying. Although this process can be reversed, it is better to start with a completely healthy animal. This sponge is extremely intolerant of suspended material in the aquarium, such as fine sand particles and detritus that can clog their inhalant siphons. Do not expose the sponge to air, as bubbles can form in the tissue when it is immersed. This will result in the death of the surrounding area.

WILL it reproduce in an aquarium?
Sometimes drops sections of body that can re-attach to substrate and form a new colony.

Haliclona molitba

Purple sponge

PROFILE

A stunning sponge that is imported fairly regularly, albeit in small numbers. It makes a wonderful display and is relatively hardy compared to other commonly available sponges. It is another species that can thrive in an aquarium with strong illumination.

WHAT size?
Grows to around 15cm in height and 20cm in diameter.

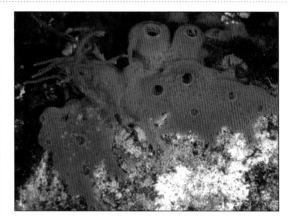

◀ *This beautiful purple sponge is often available with a more vase-like growth form than the specimen shown here. Given the correct husbandry, this sponge can thrive in the home aquarium.*

WHAT does it eat?
Feeds on very fine particulates, including bacteria and, possibly, dissolved organic material. In the aquarium, be sure to provide phytoplankton on a daily basis for the best results. Oyster eggs are another suitable diet.

WHERE is it from?
Tropical Indo-Pacific.

WHAT does it cost?
★★★☆☆
Price depends on the size of the animal.

SIMILAR SPECIES
The blue tubular sponge *(Haliclona sp.)* is similar in form but readily distinguished by its characteristic coloration (see page 17).

HOW compatible with other invertebrates?
Fairly aggressive. Secretes chemicals that may irritate or even kill animals in close proximity. It has the ability to overgrow other sessile invertebrates, but is also susceptible to damage from stinging animals.

WHAT water flow rate?
Requires strong currents.

HOW much light?
Able to withstand direct illumination, but does not contain photosynthetic algae. Can usually withstand algae that might over-grow less light-tolerant sponges.

WILL fish pose a threat?
May be nibbled by fish, including true angelfish, for whom sponges form part of their natural diet. However, this behaviour is rare due to the animal's chemical defences.

WILL it threaten fish?
This sponge should not harm fish.

WHAT to watch out for?
Specimens sometimes show a clear 'sheath' of tissue over their surface. This is often shed, revealing uniformly purple flesh underneath. However, it is not a good idea to acquire the animal until this process is complete. Colours should be rich and vibrant; seek out specimens that have settled for at least a week in the dealer's aquarium. Do not expose the sponge to air, as bubbles can form in the tissue when it is immersed. This will result in the death of the surrounding area.

WILL it reproduce in an aquarium?
Will drop sections of body material that are able to re-attach to suitable substrates.

Ianthella hasta

Purple cup sponge

PROFILE

A beautiful sponge, despite its somewhat rubbery appearance. Purple is the colour most commonly imported for the aquarium trade, but yellow and green forms are sometimes available.

WHAT size?

Usually offered for sale at 10-20cm, but can grow to double this size or more.

WHAT does it eat?

Filters fine material from the water column. Adding phytoplankton can result in rapid growth, but the flexible tissue contains photosynthetic algae that provide the majority of this animals requirements in the home aquarium.

WHERE is it from?

Tropical Indo-Pacific.

WHAT does it cost?

★★★☆☆
Price depends on size.

HOW compatible with other invertebrates?

Unlikely to threaten other sessile invertebrates, but its prodigious growth rate and large potential size mean that it can shade other animals and compete with them for available substrate.

WHAT water flow rate?

Enjoys strong water currents that help to keep its surface free from detritus and sediment.

HOW much light?

Thrives under strong illumination. T5 or metal-halide lamps are ideal.

WILL fish pose a threat?

May be nibbled by fish, including true angelfish, for whom sponges form part of their natural diet. However, this behaviour is rare due to the animal's chemical defences.

WILL it threaten fish?

Should not affect any fish species in the aquarium.

SIMILAR SPECIES

There are several species of sponge that include photosynthetic algae in their tissues. Most of these are accidental imports and many have a rubbery appearance. The ear sponge *Collospongia* is similar in colour and texture to the purple cup sponge. The chicken liver sponge (*Chondrilla* spp.) is commonly found associated with the base rock of specimen invertebrates and can grow well in reef aquariums, taking advantage of the conditions created for photosynthetic corals.

WHAT to watch out for?

The vase- or cup-shaped specimens have a tendency to collect detritus, especially where there is insufficient water flow. This must be cleared away to reduce the risks of tissue die-off or of encouraging bacterial infections to take hold.

WILL it reproduce in an aquarium?

Unlikely. It is possible to take cuttings from this species.

◁ *Although fluted shapes are more commonly encountered in the hobby,* Ianthella *spp. growth forms can depend upon the local conditions where they are collected. Fortunately, all of them are beautiful!*

Pseudaxinella lunaecharta

Orange ball sponge

Another species of sponge that is commonly acquired because of its wonderful coloration. Its aquarium hardiness will depend on whether the aquarist can provide the nutrient-poor conditions it requires, as well as sufficient suitable foods.

WHAT size?
Usually offered for sale at 10-15cm, but can grow to double this size or more.

WHAT does it eat?
Requires regular feeding with tiny particulate food, such as phytoplankton and microplankton. Oyster eggs are a suitable microplanktonic food.

WHERE is it from?
The Caribbean.

WHAT does it cost?
★☆☆☆☆
Price depends on the size of the animal.

SIMILAR SPECIES
Orange ball sponges are unlikely to be confused with any other species unless they are very small, in which case they may resemble *Tethya* – the moon sponges. However, moon sponges are more granular in appearance.

HOW compatible with other invertebrates?
This animal secretes noxious chemicals that can affect the growth of its neighbours.

WHAT water flow rate?
Enjoys strong water currents that help to keep its surface free from detritus and sediment.

HOW much light?
Intolerant of the algae that can smother them if they are sited under direct illumination. Place them in shaded areas unless you can control algal growth through water quality.

WILL fish pose a threat?
May be nibbled by fish, including true angelfish, for whom sponges form part of their natural diet. However, this behaviour is rare due to the animal's chemical defences.

▲ *Although this species can tolerate a certain amount of detritus or sand coverage, it prefers to be kept free from fouling.*

WILL it threaten fish?
Stressed or dying ball sponges can secrete toxic chemicals into the water that have an adverse effect on fish.

WHAT to watch out for?
Look for specimens with uniform coloration and no signs of encroaching algae. The margins of unhealthy individuals often appear transparent. Do not expose the sponge to air, as bubbles can form in the tissue when it is immersed. This will result in the death of the surrounding area.

WILL it reproduce in an aquarium?
Unlikely.

Ptilocaulis sp.

Orange, or red, tree sponge

PROFILE

A striking species of sponge that is commonly available in the hobby. It is often sold when fairly large and consequently can command a fairly high price due to the shipping costs involved. It can be difficult to site in the aquarium as it has a high centre of gravity and usually only a small amount of substrate attached.

WHAT size?
Forms branching colonies up to 30cm wide and 45cm tall.

WHAT does it eat?
Feeds on very fine particulates, including bacteria and, possibly, dissolved organic material. In the aquarium, be sure to provide phytoplankton on a daily basis for the best results. Oyster eggs are another suitable diet.

WHERE is it from?
The Caribbean. Similar species are found in the Indo-Pacific but imported with less regularity.

WHAT does it cost?
★★★☆☆
Price depends on the size of the animal.

HOW compatible with other invertebrates?
Not particularly aggressive as it grows up and out, rather than in an encrusting fashion. Therefore, it will not smother its neighbours.

WHAT water flow rate?
Strong currents are absolutely essential for this species and its close relatives. Some authorities state that unidirectional constant flow is essential, but it should be high-flow, low-velocity.

▼ *Select branching sponges carefully, not just for condition and vitality but also for shape. Large individuals are highly attractive but can prove difficult to anchor to rockwork, and they will not enjoy repeated handling if they continually fall over.*

HOW much light?
Can tolerate direct illumination, but best sited in areas where they will receive less light.

WILL fish pose a threat?
May be nibbled by fish, including true angelfish, for whom sponges form part of their natural diet. However, this behaviour is rare due to the animal's chemical defences.

WILL it threaten fish?
Should not harm fish.

WHAT to watch out for?
Avoid specimens with any pale patches of tissue. The margins of the branches will often show the first signs; they are an indication of tissue die-off. The colour should be uniform and vibrant. This sponge is extremely intolerant of inedible particulates in the water. Do not expose it to air, as bubbles can form in the tissue when it is immersed, resulting in damage.

WILL it reproduce in an aquarium?
Unlikely.

> #### SIMILAR SPECIES
> Members of the genera *Higginsia, Aplysina* and *Amphimedon* are similar in growth form and appearance, although some may be orange or yellow and the skin may have a rougher appearance.

Beautiful but with a sting!

Sea anemones are often the first animals to attract aquarists to a reef aquarium, not least on account of the relationship that some species share with anemonefish. However, only 10 species are known to host these fish and compared with other members of this order, they can be more challenging to maintain. Most of the anemones regularly available in the hobby contain photosynthetic algae and require relatively high levels of light and excellent water quality. However, their stinging tentacles are capable of ensnaring food particles, so most specimens benefit from regular feeding. Bear in mind the ability of most species to roam around the aquarium and the implications that this behaviour may have on other tankmates. However, in stable conditions, a settled specimen is likely to remain in the same position for years.

Price guide

★	£5 – 15
★★	£15 – 35
★★★	£35 – 50
★★★★	£50 – 80
★★★★★	£80 – 100

PROFILE

The obvious beauty of this anemone belies the problems it can cause in the home aquarium. However, given a suitable aquarium without any fish, and in the presence of its usual commensal organisms – shrimps from the genus *Periclimenes* – this animal makes a wonderful display. It has the most potent sting of the anemones that are imported for the hobby.

WHAT size?
Can reach over 30cm across the tentacles.

WHAT does it eat?
Feed regularly (every few days) with fine, particulate material, such as chopped mysis or brineshrimp, preferably enriched. Contains photosynthetic algae that require strong illumination.

WHERE is it from?
Tropical Indo-Pacific.

WHAT does it cost?
★★☆☆☆

Actinodendron glomeratum

Fire anemone

⬥ *Fire anemones are stunning and will often thrive in captivity, but they present a clear danger to other aquarium inhabitants.*

Other species of *Actinodendron* are sometimes sold and they are very similar in terms of appearance and aquarium requirements.

amounts of mucus that help it to become neutrally buoyant. It can then float around the aquarium, stinging anything in its way, hence the need to maintain it on its own. Can inflict a very painful sting on unwary aquarists. Always use gloves when cleaning or feeding.

WILL it reproduce in an aquarium?

Not reported, but bilateral fission is possible. In this case, the anemone divides along its longitudinal axis.

HOW many in one tank?
Keep singly.

HOW compatible with other invertebrates?
Will sting its neighbours and anything else that strays too close. Commensal shrimps from the genus *Periclimenes* are known to inhabit the tentacles of this anemone.

WHAT minimum size tank?
As this species grows to a large size, allow 200 litres for a single specimen.

WILL fish pose a threat?
Most fish have an inherent wariness of this anemone and ignore and avoid it.

WILL it threaten fish?
Yes. It has the potential to kill and possibly consume fish that come into contact with its tentacles.

WHAT to watch out for?
This anemone prefers to burrow into soft sandy substrates. It can sometimes secrete copious

PHYLUM CNIDARIA

Cnidaria is a name given to a group of animals that all possess stinging cells. The 10,000 species currently described include a wide diversity of radially symmetrical animals such as sea anemones, corals (both soft and stony), colonial polyps and jellies.

Bartholomea annulata

Curlicue, or corkscrew, anemone

PROFILE

Similar in appearance to its close relative, the glass rose, or triffid, anemone (*Aiptasia* sp.), the curlicue is occasionally offered for sale as a hardy alternative to other anemones in the aquarium. It can be distinguished from similar species by the corkscrew patterns of white pigment that wind their way up the tentacles, and its much larger final size.

WHAT size?
Can reach about 18cm across the tentacles.

WHAT does it eat?
Will consume particulate diets, such as mysis, brineshrimp and chopped shellfish. Does not contain photosynthetic algae. Feed daily.

WHERE is it from?
Tropical Western Atlantic-Caribbean.

WHAT does it cost?
★☆☆☆☆
Price depends on size.

SIMILAR SPECIES
Most likely to be confused with *Aiptasia*, but a number of tropical anemones may appear similar until the characteristic patterning of the tentacles is taken into consideration.

▲ *The markings in the tentacles help to distinguish this species from similar pest anemones, such as* Aiptasia.

HOW many in one tank?
Keep singly or in groups, but monitor their expansion and asexual reproduction. This anemone can proliferate when conditions are favourable.

HOW compatible with other invertebrates?
Will sting its neighbours and anything else that strays too close.

WHAT minimum size tank?
150 litres or more.

WILL fish pose a threat?
Few fish will bother this anemone. Some, for example cardinalfish, may occasionally use it as a refuge in the aquarium.

WILL it threaten fish?
Can inflict a sting, but most fish will keep a respectful distance.

WHAT to watch out for?
Ideally, this anemone should be exposed to intermittent water currents of moderate intensity. Although it will inhabit rocks, coarse gravel and coral rubble are also suitable substrates.

WILL it reproduce in an aquarium?
Able to reproduce asexually through fission, but this will depend on the conditions in the aquarium. In wild specimens, such reproduction is usually limited to the summer months.

Cerianthus spp. *Pachycerianthus* spp.

Tube anemone

PROFILE

Living in a tube comprised of your own faecal material, discarded stinging cells and mucus is not everyone's idea of a pleasant abode and yet this is what protects the tube anemones from predators. Several colour forms are available, including orange purple, black and white. This anemone has a bad reputation with aquarists, but recent studies suggest this might be ill-deserved. Small specimens might be confused with sea cucumbers

WHAT size?
Large specimens reach 45-50cm or so across the outer tentacle ring.

WHAT does it eat?
Readily accepts small chunks of almost any shellfish. Mysis and brineshrimp are also suitable. Offer a varied diet every day. These anemones do not contain photosynthetic symbionts.

WHERE is it from?
Tropical Indo-Pacific.

WHAT does it cost?
★☆☆☆☆ ★★★☆☆
Price depends largely on size and colour.

HOW many in one tank?
Can be kept in groups, but given its potential to harm fish and invertebrates, this anemone is usually maintained singly.

HOW compatible with other invertebrates?
Will sting its neighbours, but with no greater impact than pest anemone species such as *Aiptasia* (the glass rose, or triffid, anemone).

WHAT minimum size tank?
150 litres or more.

WILL fish pose a threat?
Most fish will ignore tube anemones, although some might nibble at their protective tube or commensal animals that reside there. This anemone species does not host anemonefish.

WILL it threaten fish?
Yes. It can sting fish and leave nasty-looking welts on their bodies. However, this is unusual and rarely fatal, particularly if the fish are otherwise healthy.

NO REAL THREAT

Recent studies suggest that this anemone is no worse than many commonly kept species that host anemonefish in terms of the threat it poses to aquarium inhabitants. Small specimens might be confused with sea cucumbers.

This anemone does not include fish in its natural diet, instead feeding on plankton.

WHAT to watch out for?
Avoid specimens with short tentacles as these may have been starved for prolonged periods. Allow them plenty of space in which to spread their tentacles and dig themselves into the sand.

WILL it reproduce in an aquarium?
Asexual and sexual reproduction are rare in aquarium specimens of this species.

◀ The striking colours of tube anemones mean that aquarists are willing to stock them, despite their reputation for inflicting potentially dangerous stings.

Condylactis passiflora

Atlantic anemone

This hardy anemone does not naturally play host to anemonefish, although certain species of the latter may adopt it in the absence of anything they deem more suitable. Coloration is variable, from almost snow-white to deep purple, with various shades in between.

WHAT size?
Large specimens reach 15-20cm or so across the tentacle disc.

WHAT does it eat?
Offer shellfish every few days. Chunks of fish, mussel, cockle or clam are ideal. The anemone contains symbiotic algae.

WHERE is it from?
Tropical Atlantic and Caribbean.

WHAT does it cost?
★☆☆☆☆ ★★☆☆☆
Price depends on size and colour. Purple specimens are often more expensive than drabber individuals.

SIMILAR SPECIES
Condylactis gigantea is similar in terms of husbandry but grows larger, even in the home aquarium.

▲ *If there is such an animal as a 'beginner's anemone' this could be it. Attractive and one of the easier species to maintain in the home aquarium.*

HOW many in one tank?
Can be kept singly or in groups.

HOW compatible with other invertebrates?
Will sting its neighbours and anything that strays too close.

WHAT minimum size tank?
150 litres or more.

WILL fish pose a threat?
Some larger angelfish may nip at this species, but most fish are innately wary of an anemone's sting.

WILL it threaten fish?
Yes. It can sting fish residents, but such incidents are rare. Most fish species easily avoid the large tentacles, but accidents can and do happen.

WHAT to watch out for?
Avoid deflated specimens. Healthy individuals will have a substantial appearance, with a fully inflated column and tentacles.

WILL it reproduce in an aquarium?
Asexual and sexual reproduction are rare in aquarium specimens of this species.

Entacmaea quadricolor

Bubble, or Gelam, anemone

PROFILE

Although it is one of the hardier choices for the marine aquarist, the bubble, or bubbletip, anemone still demands excellent water quality and aquarium stability. It is renowned for taking a while to settle into its preferred position in the aquarium, but usually stretches its foot down into crevices in the rockwork and opens the oral disc into open water.

WHAT size?
Large specimens reach 60cm or so across the tentacle disc.

WHAT does it eat?
Contains zooxanthellae and will survive with only sporadic feedings (every 7-10 days). Some people advocate more regular offerings of shellfish and whole fish, such as silversides, every 2-3 days.

WHERE is it from?
Tropical Indo-Pacific, including the Red Sea.

WHAT does it cost?
★★☆☆☆   ★★★★★
Price depends on size and colour. Red specimens always demand a high premium and are commonly known as rose anemones.

▶ *An orange skunk anemonefish (Amphiprion sandaracinos) finds sanctuary and a perching point in the 'arms' of a bubble anemone.*

HOW many in one tank?
Can be kept in groups, if you can tolerate the problems caused by a number of wandering stinging specimens in the same aquarium.

HOW compatible with other invertebrates?
Problematic because it has stinging tentacles and will frequently roam the aquarium before settling on a suitable position. Likely to upend specimen corals or sit directly on top of them with varying implications, depending on the hardiness of the victim.

WHAT minimum size tank?
200 litres should accommodate even large specimens, but be aware of the roaming behaviour of this species.

WILL fish pose a threat?
Fish tend to target only damaged or stressed specimens. They will peck at the internal organs of the invertebrate through the gaping oral opening.

WILL it threaten fish?
Yes, can sting and sometimes

SIMILAR SPECIES
Sometimes confused with some of the less spectacular specimens of *Heteractis magnifica* and *Macrodactyla doreensis*. The former generally has quite uniformly cylindrical tentacles ending in a rounded tip. The latter usually has a red base, tapering tentacles and prefers sandy substrates.

consume fish but this is rare. It is especially unlikely if the anemone is playing host to a pair of anemonefish who will drive other fish away. The maroon anemonefish *(Premnas biaculeatus)* is only found in bubbletip anemones and will often move into one as soon as it is stocked into the aquarium.

WHAT to watch out for?
Avoid very pale specimens, as they may have lost a large proportion of their symbiotic algae. Pay particular attention to the base. Do not buy deflated or damaged specimens.

WILL it reproduce in an aquarium?
Asexual reproduction is relatively common and may be a result of local competition. Specimens that clone themselves easily will often divide regularly – sometimes as often as every few weeks. Spawning has been recorded in home aquariums.

Heteractis crispa

Leathery, or crispa, anemone

H.crispa is a beautiful species of host anemone that can prove hardy once established in the home aquarium. Although capable of stinging, the greatest threat presented by any anemone is the pollution it can cause if it dies.

◀ *The long, tapering tentacles of* H. crispa *distinguish it from* H. malu. *Ideally, the tentacles should feel slightly tacky to the touch.*

WHAT size?
Large specimens reach 60cm or so across the tentacle disc. Allow plenty of space for this anemone to expand.

WHAT does it eat?
Contains photosynthetic symbionts, but offer it chunks of alternative meaty food every few days or so. Mussel, prawn, shrimp and similar shellfish are suitable. Commonly available with no zooxanthellae ('white malu') i.e. bleached. Such specimens need feeding more regularly to buy them time until they regain their photosynthetic pigments. It is best to avoid white specimens at all costs.

WHERE is it from?
Tropical Indo-Pacific.

WHAT does it cost?
★★☆☆☆ ★★★☆☆
Price depends on size.

HOW many in one tank?
Best kept singly due to its large ultimate size.

HOW compatible with other invertebrates?
Stinging cells are capable of damaging neighbouring corals.

WHAT minimum size tank?
Large specimens will require systems of 450 litres or more.

WILL fish pose a threat?
Fish tend to target only damaged or stressed specimens. They will peck at the invertebrate's internal organs through the gaping oral opening.

WILL it threaten fish?
Yes. Can sting and sometimes consume fish, but this is rare. Most popular aquarium fish seem to have an innate recognition of anemones and their potential to do harm.

WHAT to watch out for?
Avoid white (bleached), yellow (artificially coloured) and deflated specimens. Choose specimens that have been in the retailer's aquarium for at least two weeks and show no signs of stress or damage.

WILL it reproduce in an aquarium?
Asexual reproduction has been reported but is rare. There are several anecdotal references to spawning events occurring in home aquariums. Anemones are dioecious, meaning that separate sexes occur.

SIMILAR SPECIES
Often confused with *H. malu*, as the most commonly available colour morphs are very similar shades of beige. However, *H. crispa* has relatively long, tapering tentacles and a leathery texture to the column. The tentacles of *H. malu* are fatter in the middle than at the base or tip, and much shorter than in *H. crispa*. Several anemonefish use this species as a host, including common clownfish (*Amphiprion ocellaris*).

Heteractis magnifica

Magnificent, or ritteri, anemone

PROFILE

A beautiful species that can be difficult to care for in the home aquarium. It usually does best under strong metal-halide illumination, often with low Kelvin ratings (6,500K or less). Most specimens do not ship well; choose individuals by their appearance. Fully inflated, settled specimens are ideal, but this anemone is best left to experienced aquarists.

WHAT size?
Large specimens reach 90-100cm or so across the tentacle disc.

WHAT does it eat?
Contains photosynthetic symbionts but offer chunks of alternative meaty food every few days or so. Mussel, prawn, shrimp and similar shellfish are suitable.

WHERE is it from?
Tropical Indo-Pacific, including the Red Sea.

WHAT does it cost?
★★☆☆☆   ★★★★★
Price depends on size, source and colour.

▶ *The purple base of this anemonefish host clearly identifies it as* H. magnifica, *although other colour forms do occur. Note the cylindrical tentacles with slightly rounded ends.*

HOW many in one tank?
Can be kept in numbers, but this species grows very large and is highly mobile when first introduced.

HOW compatible with other invertebrates?
The potent sting of this species, coupled with its tendency to roam, means that it is not recommended for aquariums containing sensitive sessile invertebrates. This includes many small- and large-polyp stony corals. You will have to move corals around the anemone when it settles and allow for its prodigious growth.

WHAT minimum size tank?
Realistically, to house this species long term, provide a very large aquarium of 1200 litres or more.

WILL fish pose a threat?
Most fish will ignore anemones unless the anemone is stressed or damaged. Butterflyfish and dwarf angelfish may peck at such specimens, often targeting

SIMILAR SPECIES
Drabber specimens closely resemble bubble anemones (*Entacmaea quadricolor*), but lack the bubble tips. The tentacles of *H. magnifica* are usually uniformly cylindrical along their length, whereas similar species tend to taper to a pointed tip.

their internal tissues, which they access through the gaping oral opening.

WILL it threaten fish?
Yes. Will sting fish that contact the tentacles. Anemonefish are the obvious exception. Percula (*Amphiprion percula*) and common clownfish (*Amphiprion ocellaris*) are particularly fond of this anemone, but it plays host to at least half a dozen additional species.

WHAT to watch out for?
Avoid specimens that appear deflated or have gaping mouths. Check the base of the stalk (the foot) for signs of tearing. Only choose specimens that have recovered fully from shipping, or seek an alternative host for your anemonefish.

WILL it reproduce in an aquarium?
Asexual reproduction has been reported but is rare. It is likely that larger, well-settled specimens will spawn, but records are scarce.

Macrodactyla doreensis

Sand, or long-tentacled, anemone

PROFILE

Perceived by some people to be amongst the hardiest of host anemone species and by others as one of the most difficult. Excellent water quality and a generous depth of sand or fine gravel are no guarantee of success with this species. In the home aquarium, success is likely to depend on the individual specimen, rather than the aquarist's best efforts.

WHAT size?
Large specimens reach 45-50cm or so across the tentacle disc.

WHAT does it eat?
Offer shellfish every few days. Choose chunks that are too large to be comfortably consumed by fish residents.

WHERE is it from?
Tropical Indo-Pacific.

WHAT does it cost?
★★☆☆☆
★★★★☆
Price depends on size and colour. Purple and red specimens often command a premium.

HOW many in one tank?
Best kept singly, as this species can grow large and will often refuse to dig itself into the substrate. This often results in the anemone drifting in the water currents, stinging anything that it comes into contact with.

HOW compatible with other invertebrates?
Will sting its neighbours and anything that strays too close.

WHAT minimum size tank?
300 litres, but larger specimens will dominate such systems.

WILL fish pose a threat?
Fish tend to target only damaged or stressed specimens. This anemone has potent stinging cells that make it feel slightly sticky to the touch. This tends to deter most fish from nibbling at it.

▼ *Healthy sand anemones will spread their tentacle crown over the substrate and withdraw quickly into the sand when disturbed.*

WILL it threaten fish?
Yes. It can sting fish residents but such incidents are rare. Try to avoid keeping fish that are slow moving, apparently clumsy or nervous, as they can either blunder into anemones or have difficulty extricating themselves should they make contact with the tentacles.

WHAT to watch out for?
Avoid specimens that have not dug themselves into the substrate in the retailer's aquarium (where substrate has been provided). Avoid obviously damaged, deflated or excessively pale individuals.

WILL it reproduce in an aquarium?
Asexual and sexual reproduction are rare in aquarium specimens.

SIMILAR SPECIES

Small individuals can be confused with the beaded anemone, *Heteractis aurora*. It is a hosting species that often has a red base, but does not grow much larger than 20cm. Its ornate tentacles with lateral extensions of tissue give it a very beautiful appearance. *H. aurora* is often referred to as a short-tentacled sand anemone, as it too is found in soft substrates. It plays host to a variety of anemonefish, particularly juvenile specimens.

Stichodactylus haddoni

Carpet, or giant carpet, anemone

PROFILE

One of the largest species of sea anemone available to hobbyists. The densely packed short tentacles endow the animal with a powerful battery of stinging cells that are extremely sticky to the touch and capable of inflicting nasty injuries, both on the aquarist and aquarium inhabitants. Best housed in a large aquarium.

◀ *Purple is just one of the stunning natural colour varieties of this species, but the more unusual it is, the higher the price! Blue, purple and red are amongst the most desirable.*

WHAT size?
Large specimens reach 50-60cm or so across the tentacle disc.

WHAT does it eat?
Offer shellfish as often as every day. Chunks of fish, mussel, cockle or clam are ideal. The anemone contains symbiotic algae.

WHERE is it from?
Tropical Indo-Pacific. Many collected specimens, particularly those with bright green, purple and blue coloration, are exported through Sri Lanka.

WHAT does it cost?
★★☆☆☆   ★★★★☆
Price depends on size and colour. Brown or dull-green individuals are the least expensive. Ignore white specimens entirely, as they have lost their photosynthetic pigments.

HOW many in one tank?
Best kept singly due to its potential maximum size.

HOW compatible with other invertebrates?
Will sting its neighbours and anything else that strays too close.

WHAT minimum size tank?
450 litres is sufficient for a smaller specimen, but a large individual will dominate a system of this volume.

WILL fish pose a threat?
Some brave fish will peck at this anemone but only the most hardy will do it twice!

WILL it threaten fish?
Yes. It can sting and capture fish. Keeping a pair of anemonefish can sometimes reduce this risk, as they drive away tankmates that get too close, but almost any resident fish is at risk from this invertebrate.

WHAT to watch out for?
Avoid deflated specimens or those with a gaping mouth. Healthy individuals will withdraw quickly when touched and have very sticky tentacles. Handle them with gloves to prevent damaging the anemone or being injured yourself.

WILL it reproduce in an aquarium?
Asexual and sexual reproduction are rare in aquarium specimens of this species.

SIMILAR SPECIES

At least two other *Stichodactyla* species make their way into the hobby. Their husbandry is similar to that of *S. haddoni*. Another anemone, *Cryptodendrum adhaesivum*, the pizza anemone, is sometimes imported. Although beautiful, it can prove very difficult to keep long term. All these species have small, densely packed tentacles, but only around the outer margin of the oral disc.

Elegant and varied

Soft corals and colonial polyps are represented by around 2,500 described species. The animals contained in this assemblage are highly diverse in appearance; some resemble small sea anemones, whereas others can grow into huge, candelabra-like structures. The majority of species of interest to aquarists contain symbiotic algae and as a result enjoy good illumination, but in general this does not have to be as intense as that required for stony corals. Soft corals and colonial polyps can influence their neighbours, not only by directly stinging their competitors, but also by releasing noxious compounds from their tissues. To prevent a single specimen from dominating an aquarium, it is necessary to introduce some means of removing these chemicals from the water. A protein skimmer will usually accomplish this successfully.

Price guide

★	£5 – 10
★★	£10 – 20
★★★	£20 – 30
★★★★	£30 – 50
★★★★★	£50 – 70

Capnella spp.

Asparagus tip coral

HOW compatible with other invertebrates?

Few invertebrates represent a threat to this coral.

WHAT water flow rate?

Moderate flow is essential. This coral is not as flexible as medusa or colt coral and therefore unable to withstand very strong currents.

HOW much light?

Requires strong illumination. Best maintained under T5 or metal-halide lighting.

WILL fish pose a threat?

Some fish might peck at the polyps of this branching soft coral, but such experiences are uncommon amongst reef-safe fish in the aquarium.

WILL it threaten fish?

No, harmless to fish.

WHAT to watch out for?

This coral can appear quite uninspiring when first imported and needs an extended period to acclimatise and settle down. The same applies once it has been introduced into the aquarium.

WILL it reproduce in an aquarium?

Drops small clusters of polyps that adhere to the substrate and grow. Sexual reproduction is rare. Easy to propagate.

▼ *Subtle differences in growth form and polyp arrangement aid identification.*

◀ *Soft corals can harbour harmless commensal animals. In this image, a brittle starfish resides on the stalk of the coral.*

SIMILAR SPECIES

Easily confused with colt coral (*Alcyonium* spp.) and bush corals (*Litophyton*, *Nephthea* and *Lemnalia* spp.), but distinguishable by its characteristic polyp form that resembles the tip of asparagus.

Cespitularia sp.

African blue coral

This stunning coral is closely related to the pulse corals (*Xenia* and *Heteroxenia*). Like them, it can be temperamental in the aquarium. Some specimens thrive and grow, whereas others simply waste away.

WHAT size?
Has the potential to grow quite large, but aquarium specimens seldom achieve more than 30cm in diameter.

WHAT does it eat?
This coral contains photosynthetic symbionts, but may also remove microplankton from the water column.

WHERE is it from?
Tropical Indo-Pacific.

WHAT does it cost?
★★☆☆   ★★★★☆
Expensive due to shipping charges, unless you can obtain a cutting from another hobbyist or dealer.

SIMILAR SPECIES
There are several species contained in the genus *Cespitularia*. Some are as beautiful as the African blue, whereas others are drabber but still generally stunning in overall appearance.

▲ *Under strong illumination the coral retains its fluorescent blue pigment.*

HOW compatible with other invertebrates?
An aggressive species that overwhelms its neighbours through a combination of prodigious growth and chemicals (terpenoids) exuded from its tissue.

WHAT water flow rate?
Strong, indirect flow is essential for the best chances of success with this coral.

HOW much light?
Requires strong illumination to satisfy its nutritional requirements and to assist the retention of the beautiful blue pigments.

WILL fish pose a threat?
Few fish show any interest in this coral, possibly due to the fact that it can secrete noxious substances.

WILL it threaten fish?
No, harmless to fish.

WHAT to watch out for?
Specimens should appear fully inflated and, ideally, bright blue under strong lighting. Brown specimens can regain their coloration under strong illumination. Check for signs of growth around the base of the coral; fleshy extensions can indicate new growth since importation and is a sign of a colony in good condition.

WILL it reproduce in an aquarium?
Most reproduction is asexual, but spawning events cannot be ruled out in healthy, thriving colonies.

Cladiella spp.

Cauliflower coral

PROFILE

Cauliflower coral, so named for its lumpy appearance when the polyps are retracted, grows closer to the substrate than some other members of the genus. Its encrusting growth rapidly colonises new substrate and it will require plenty of space in which to grow unless cuttings are taken.

WHAT size?
Encrusting specimens will colonise almost any vacant hard surface in the aquarium, potentially reaching over 60cm in diameter, even in a system of modest proportions.

WHAT does it eat?
Contains symbiotic algae that satisfy the majority of its nutritional requirements, but will also trap small particulate food such as microplankton and brineshrimp larvae.

WHERE is it from?
Tropical Indo-Pacific.

WHAT does it cost?
★★☆☆☆
★★★★☆
This price range should secure a specimen of cauliflower coral.

HOW compatible with other invertebrates?
This is a fast-growing, aggressive coral that secretes some noxious compounds designed to repel its neighbours. Take care that it does not begin to dominate an aquarium, as it is able to retard the growth of its tankmates.

WHAT water flow rate?
Cauliflower coral will tolerate a range of flow strengths, from almost completely slack water to quite strong, indirect flow.

HOW much light?
This coral can be kept under T8, T5 or metal-halide lighting.

WILL fish pose a threat?
Some fish will peck at the coral's polyps. This includes some dwarf angelfish, although this is unusual. Most reef-safe fish will ignore it completely.

SIMILAR SPECIES
Cauliflower corals are generally white with brown polyps, but other species of soft coral, including the colt or pussey coral (*Alcyonium* spp.), are often referred to as cladiella due to the similar polyp form.

WILL it threaten fish?
No, harmless to fish.

WHAT to watch out for?
This coral is often predated by *Tritoniopsis elegans*, a nudibranch predator commonly imported by accident with soft coral specimens. A single specimen of this sea slug can cause massive damage to the coral. Inspect a potential purchase for contracted polyps or powdery deposits around the base.

WILL it reproduce in an aquarium?
The cauliflower coral will drop polyps of its own accord and is easy to propagate by taking cuttings. Sexual reproduction is rare.

◀ *White tissue underlying brown polyps is characteristic of many forms of* Cladiella.

Dendronephthya spp.

Prickly soft coral

PROFILE

This beautiful coral does not contain photosynthetic pigments and must therefore be classed as difficult. However, many aquarists report successes now that suitable invertebrate foods have become widely available.

WHAT size?
Wild specimens can expand to well over 60cm tall and wide. Aquarium specimens seldom thrive to this extent.

WHAT does it eat?
Does not derive energy from photosynthetic symbionts. Provide regular offerings of phytoplankton and a variety of other very fine particulate foods. Reducing mechanical filtration and protein skimming have also been linked with success. It has been suggested that this coral feeds by taking up dissolved organic compounds directly from the water.

WHERE is it from?
Tropical Indo-Pacific.

WHAT does it cost?
★★★☆☆   ★★★★★
Specimens can be very expensive, especially brightly coloured individuals.

HOW compatible with other invertebrates?
Presents no threat to other corals, but be aware of the implications for water quality of adding the copious amounts of food that this species seems to need.

WHAT water flow rate?
Prefers moderate to strong flow.

HOW much light?
No special requirements. May be reluctant to expand during the day, but not strictly speaking nocturnal.

WILL fish pose a threat?
Few reef-compatible fish show any interest in this soft coral. It contains sharp sclerites (rigid calcium carbonate components) that help to defend its soft tissues.

WILL it threaten fish?
No, harmless to fish.

WHAT to watch out for?
Avoid deflated specimens. This is typical advice for all corals, but in non-photosynthetic species it does not necessarily signify ill-health. However, unless the coral is prepared to come out during the day, assessing its overall condition is practically impossible.

WILL it reproduce in an aquarium?
Will drop small fragments that usually thrive in the aquarium. Cuttings can be taken from well-settled individuals.

▲ *Beautiful colours characterise corals from the genus* Dendronephthya, *tempting aquarists who are unaware of their specific requirements.*

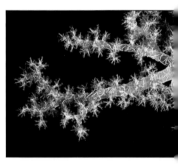

▲ *Keep* Dendronephthya *in an aquarium where it can be fed regularly.*

SIMILAR SPECIES
Neospongodes is a photosynthetic coral with a passing resemblance to *Dendronephthya,* but other non-photosynthetic species, such as *Scleronephthya,* are closest in appearance.

Klyxum spp. ('*Alcyonium*')

Colt, or pussey, coral

PROFILE

In an aquarium that suits its requirements, this popular soft coral can achieve prodigious growth. It does not contain as many supportive elements as many other soft corals and can thus show great differences in size between the expanded and contracted phases.

WHAT size?
Can grow to fill an aquarium with its branching form. Wild specimens can measure over one metre across.

WHAT does it eat?
Primarily photosynthetic in the marine aquarium, but may also consume very fine particulate food.

WHERE is it from?
Tropical Indo-Pacific.

WHAT does it cost?
★★☆☆☆
This coral needs to be shipped in plenty of water, so it is rarely available for less than the price rating given here, although small cultured specimens are more modestly priced. Large colonies can be significantly more expensive.

▶ *Tall, branching growth is seen in most specimens of* Klyxum *spp. It grows very quickly when the conditions are suitable.*

HOW compatible with other invertebrates?
Can quickly overshadow its neighbours, depriving them of the light they need to survive.

WHAT water flow rate?
This coral will tolerate a wide range of currents provided they are not too harsh. It will grow well with very little flow, but this is unlikely to benefit the other aquarium inhabitants.

HOW much light?
Moderate to strong illumination. Will thrive under strong T8, T5 or metal-halide lighting of sufficient intensity.

SIMILAR SPECIES
Cladiella spp. are very similar, particularly those with highly branching growth. Identifying many soft corals correctly, even to genus level, can require dissection of their tissues and is therefore almost impossible for most aquarists to achieve.

WILL fish pose a threat?
Some fish will peck at this coral, causing it to remain contracted. Few commonly maintained reef aquarium fish will be a problem.

WILL it threaten fish?
No, harmless to fish.

WHAT to watch out for?
Avoid contracted specimens. This coral is often predated by *Tritoniopsis elegans,* a nudibranch predator commonly imported by accident with soft coral specimens.

WILL it reproduce in an aquarium?
Forms daughter colonies through budding of the main column or shedding of fingerlike extensions of tissue. Easy to propagate.

Lobophytum spp.

Cathedral coral

A stunning coral with two types of polyp. Short brown ones give it a furry appearance, broken only by the much larger snow-white polyps that endow this coral with its wonderful looks.

WHAT size?
Large specimens may reach over 45cm in a home aquarium, but these are dwarfed by wild specimens that can grow to more than one metre across.

WHAT does it eat?
Contains photosynthetic pigments, but also consumes very fine particulate food.

WHERE is it from?
Tropical Indo-Pacific.

WHAT does it cost?
★★☆☆☆   ★★★★☆
Most specimens offered for sale will be in the above price range, depending on size.

SIMILAR SPECIES
Other *Lobophytum* species, often known as 'knobbly leather corals' are available. Their husbandry is similar to the species shown here, but they may resemble toadstool leather corals (*Sarcophyton* spp.). Some specimens have green or yellow coloration.

▲ *White polyps contrasted with a darker base make this a popular coral.*

HOW compatible with other invertebrates?
This coral presents little threat to any of its tankmates, but does exude chemicals from its tissue (terpenoids) and can therefore have an adverse effect on the growth of other corals.

WHAT water flow rate?
Moderate to very high. The coral secretes its waste into a thin film of mucus that must be shed regularly. Water movement around the coral enables it to do this.

HOW much light?
Thrives under strong T5 or metal-halide lighting of sufficient intensity.

WILL fish pose a threat?
Very few species of fish will attack the flesh of this coral, but some, particularly dwarf angelfish of the genus *Centropyge*, may nip at its larger polyps.

WILL it threaten fish?
No. This is a benign species of coral.

WHAT to watch out for?
Avoid individuals showing yellow patches anywhere on the upper surface, as these can indicate bacterial infections incurred through the rigours of shipping. Specimens may frustrate their owners by refusing to expand their white polyps, despite appearing otherwise healthy and growing at this time. The reasons for this are unknown.

WILL it reproduce in an aquarium?
This coral rarely drops daughter colonies, but is fairly easy to propagate by cuttings.

Neospongodes spp. *(Stereonephthya* spp.*)*

Bush coral

PROFILE

A beautiful soft coral that resembles some of the attractive yet difficult non-photosynthetic soft corals, such as those belonging to the genus *Dendronephthya*. Yellow and red forms are the most sought after, but the standard brown variant is still highly attractive.

WHAT size?
Wild colonies grow quite large, but aquarium specimens tend to remain more modestly proportioned, reaching no more than 60cm in height and 30cm wide.

WHAT does it eat?
This coral contains photosynthetic pigments, but seems to benefit from regular feedings with fine particulate invertebrate foods and phytoplankton.

WHERE is it from?
Tropical Indo-Pacific.

WHAT does it cost?
★★☆☆☆   ★★★☆☆
Price depends on colour and size. Red specimens tend to be the most expensive, followed by yellow. Brown specimens are the cheapest.

▶ *This attractive coral will not thrive unless it has access to food. Its supportive sclerites are visible in the tissue and make it easy to identify.*

HOW compatible with other invertebrates?
When large and fully expanded, it shades a significant area of the aquarium. It is not directly aggressive and appears to lack many of the chemical defences commonly found in other soft corals.

WHAT water flow rate?
Prefers moderate to strong flow.

HOW much light?
Good to strong illumination provided by T5 or metal-halide lighting.

WILL fish pose a threat?
The odd rogue element always exists in a reef aquarium, but most reef-compatible fish will ignore this coral.

SIMILAR SPECIES
Similar to photosynthetic species, such as *Litophyton*, *Nephthea* and *Capnella*, some of which are also known as bush corals. *Neospongodes* is usually readily identified by the highly visible sclerites in the tissue. Could also be confused with non-photosynthetic corals, such as *Scleronephthya* or *Dendronephthya*.

WILL it threaten fish?
No, harmless to fish.

WHAT to watch out for?
Avoid deflated specimens or those with exposed calcareous sclerites (the rigid calcium carbonate components that assist soft corals with their rigidity).

WILL it reproduce in an aquarium?
Will drop small fragments that usually thrive in the aquarium. Cuttings can be taken from well-settled individuals.

Nephthyigorgia spp.

Red chilli soft coral

A beautiful species of soft coral whose deep red base colour contrasts with white polyps. It does not photosynthesise and is unfortunately often bought by aquarists unable to provide for its long-term needs.

WHAT size?
Can reach over 30cm across, but most aquarium specimens are a more modest size.

WHAT does it eat?
This coral requires feeding with small, particulate zooplankton substitutes, such as cyclops, brineshrimp larvae and a selection of the proprietary foods that may contain items such as rotifers.

WHERE is it from?
Tropical Indo-Pacific.

WHAT does it cost?
★★☆☆☆ ★★★☆☆
Inexpensive. Cost depends entirely on size. A high-end price should buy you a large lump of coral!

SIMILAR SPECIES
Corals from the genus *Minabea* are similar in appearance, but tend to lack the white polyps present in *Nephthyigorgia*.

HOW compatible with other invertebrates?
An aggressive species that overwhelms its neighbours through a combination of a prodigious growth rate and chemicals exuded from its tissue (terpenoids).

WHAT water flow rate?
Strong, indirect flow is essential for the best chance of success with this coral.

▼ *This nocturnal species usually only opens its polyps when the aquarium is in darkness. Feed it at this time.*

HOW much light?
Independent of light, but strong illumination may cause the growth of fouling algae on the coral's surface. Placing it in shaded areas or even fixing it upside down can help to counteract this.

WILL fish pose a threat?
Some fish may target the animal's white polyps when they are expanded. However, the coral tends to open its polyps at night, meaning that most fish will leave the coral alone entirely.

WILL it threaten fish?
No.

WHAT to watch out for?
Selecting a specimen of coral that is usually more likely to open at night can be difficult, as it will often appear deflated in a dealer's tank. Ideally, specimens should be free of algae and relatively newly imported so that they have not experienced a prolonged period of starvation. Of course, they should be well acclimatised.

WILL it reproduce in an aquarium?
Has the potential to reproduce, but reports of this are rare.

Rhytisma fulvum

Yellow encrusting leather coral

PROFILE

Similar in appearance to star polyps from the genus *Briareum,* this soft coral is often overlooked by aquarists as it is hardly as spectacular as some of its relatives. However, if you monitor its growth closely, it can make a wonderful display as it spreads over rockwork and glass, giving the aquarium a mature and natural appearance.

WHAT size?
There are no upper limits to the extent to which this coral will spread, except the amount of available substrate in the aquarium.

WHAT does it eat?
This coral is believed to feed on dissolved organic substances, which are often plentiful in the marine aquarium. However, it also contains zooxanthellae and enjoys bright lighting.

WHERE is it from?
Tropical Indo-Pacific.

WHAT does it cost?
★★☆☆☆   ★★★☆☆
An inexpensive coral that is sporadically available as cultured frags.

▲ *Trim this coral regularly or give it room to colonise bare substrate.*

HOW compatible with other invertebrates?
Few invertebrates represent a threat to this coral.

WHAT water flow rate?
This species requires a moderate to strong flow.

HOW much light?
Will survive under T8 lamps, but really thrives under T5 or metal-halide lamps.

WILL fish pose a threat?
Few fish attempt to nip at this coral.

WILL it threaten fish?
No, harmless to fish.

WHAT to watch out for?
Make sure that you observe the coral with its polyps fully extended. It does not like to be overgrown by algae.

The prodigious growth rate of this species and its ability to compete with other reef dwellers mean that it can form large masses and become a nuisance if its growth is allowed to proceed unchecked.

WILL it reproduce in an aquarium?
Primarily asexual.

SIMILAR SPECIES
Could be easily confused with star polyps from the genus *Briareum,* but the polyps are generally smaller and characteristically yellow-brown.

Sarcophyton spp.
Toadstool leather coral

PROFILE

A hardy and popular species of soft coral that usually thrives in the aquarium. It is easy to propagate by cuttings, and cultured specimens are easy to source. They make excellent additions to the reef aquarium.

WHAT size?
Large specimens may reach over 60cm in diameter, dominating the aquarium.

WHAT does it eat?
Contains photosynthetic pigments, but also consumes very fine particulate food.

WHERE is it from?
Tropical Indo-Pacific.

WHAT does it cost?
★★☆☆☆ ★★★★★
Price depends on size, origin and coloration.

SIMILAR SPECIES
Cultured specimens of the yellow-coloured sulphur toadstool (*Sarcophyton elegans*) are widely available but expensive. Other species are available in a variety of different polyp colours; white, yellow or green are the most popular.

HOW compatible with other invertebrates?
Presents few problems, but the terpenoids contained in their tissues can inhibit the growth of other tankmates. This can become a particular problem if you do not take steps to remove the terpenoids, such as using efficient protein skimming.

WHAT water flow rate?
High to very high. These soft corals secrete mucus that contains their waste. Strong water currents enable the coral to shed this mucus easily. Without them, the polyps might remain retracted and algae can begin to grow on top of them, thereby preventing them from receiving light.

HOW much light?
Good to strong illumination. Will thrive under strong T8, T5 or metal-halide lighting.

WILL fish pose a threat?
Due to this animal's strong chemical defences, few, if any, fish will present a problem. It will even be ignored by many species of butterflyfish.

WILL it threaten fish?
Should not harm any fish directly,

▲ *These hardy and beautiful soft corals can grow very large, even in smaller aquariums. Fortunately, it is possible to take cuttings, which will help to keep their rapid growth in check.*

but take care if a specimen is damaged or stressed in the aquarium, as its defensive secretions are toxic to fish.

WHAT to watch out for?
Immediately after introduction, the polyps may expand temporarily, only for the coral to close down for an indeterminate period that may last hours, days or weeks. This appears to be perfectly normal, providing the coral does not appear to be disintegrating.

WILL it reproduce in an aquarium?
Yes. Can reproduce by budding daughter colonies from the main stem.

Sarcophyton glaucum

Long-tentacle toadstool leather coral

PROFILE

Although hardy, this leather coral is more temperamental than its short-tentacled relatives. Neither does it ship as well, although cultured specimens are widely available. Acquiring these might not guarantee success, but they are usually more consistent in a home aquarium than their wild-collected brethren.

WHAT size?

Large specimens may reach over 60cm and provide a stunning centrepiece.

WHAT does it eat?

Contains photosynthetic pigments, but also consumes very fine particulate food.

WHERE is it from?

Tropical Indo-Pacific.

WHAT does it cost?

★★★☆☆ ★★★★★

Price depends on size, origin and coloration. Cultured specimens are available and prove hardier than wild-collected individuals. They are more expensive than wild specimens.

▶ *The long-tentacle leather coral has a more delicate appearance than other toadstool leather corals. It is a stunning animal with a rapid growth rate where aquarium conditions are favourable.*

HOW compatible with other invertebrates?

Presents few problems, but the terpenoids in their tissues can inhibit the growth or well-being of stony corals.

WHAT water flow rate?

High to very high. Enjoys very strong currents but not aimed directly on their tissue, which is more delicate than that of many other leather coral species. Water flow helps remove detritus and algae that may foul the organism and lead to bacterial infections.

HOW much light?

Good to strong illumination. Will thrive under strong T5 or metal-halide lighting of sufficient intensity.

SIMILAR SPECIES

Other toadstool leather corals such as *S. elegans* and *S. ehrenbergi* are similar, but have a more robust appearance. The crown of the long-tentacle leather coral can appear frilly and is much more delicate than that of other members of this genus.

WILL fish pose a threat?

Few fish deemed to be reef-safe will attack this soft coral due to its well-developed chemical defences.

WILL it threaten fish?

Should not harm any fish directly, but take care if a specimen is damaged or stressed in the aquarium, as its defensive secretions are toxic to fish.

WHAT to watch out for?

The coral should show no signs of discoloration and should be expanded, with the polyps stretched. Unlike its close relative *Sarcophyton ehrenbergi,* it does not remain retracted for prolonged periods when first introduced to its new home.

WILL it reproduce in an aquarium?

Yes. Can reproduce by budding daughter colonies from the main stem.

Sinularia sp.
Medusa coral

PROFILE

A very popular soft coral that is slimy to the touch. It does not ship easily, so allow it plenty of time to recover in the dealer's aquarium before buying it. Fortunately, aquarium- and tropical-cultured specimens are commonly available, so sourcing a well-settled, hardy individual should not be difficult.

WHAT size?
Can grow to fill an aquarium with its branching form. Wild specimens can measure over one metre across.

WHAT does it eat?
Primarily photosynthetic in the marine aquarium, but may also consume very fine particulate food.

WHERE is it from?
Tropical Indo-Pacific.

WHAT does it cost?
★★☆☆☆
★★★★☆
Colonies can be expensive due to propagation or shipping costs. Price depends on size.

HOW compatible with other invertebrates?
Grows large enough to cast a sizeable shadow that will restrict light to those corals that fall under its branching arms.

WHAT water flow rate?
Tolerates moderate to strong currents. As it lacks the internal skeletal material to give it a great degree of rigidity, its growth form will depend on the amount of water flow directed towards it. Specimens in strong flow grow short and broad, whereas in less strong currents they will often become very tall.

HOW much light?
Moderate to strong illumination. Will thrive under strong T8, T5 or metal-halide lighting of sufficient intensity. Some propagated forms may survive and grow under as few as two T8 lamps.

WILL fish pose a threat?
Some fish will occasionally nip at medusa corals, but such incidents are rare.

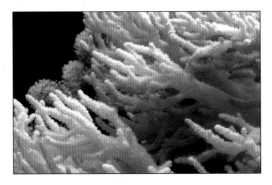

SIMILAR SPECIES
Similar to pussey coral (*Klyxum* spp.) and many *Sinularia* species. The former is usually much darker brown in colour and has larger polyps. Other *Sinularia* spp. can have a leathery texture and are more rigid in form, especially when contracted.

WILL it threaten fish?
No, harmless to fish.

WHAT to watch out for?
Avoid specimens that are mostly white or have patches that appear paler than the rest of the tissue, as they have lost photosynthetic pigments through stress. This coral is susceptible to bacterial infections that manifest themselves as black slimy deposits on the tips of the fingerlike projections.

WILL it reproduce in an aquarium?
Forms daughter colonies through budding of the main column or by shedding fingerlike extensions of tissue. Can be easily propagated by taking cuttings.

◀ The fingerlike projections of tissue in the medusa coral are often pale pink or yellow. Avoid white specimens, as these have expelled their symbiotic pigments, possibly as a result of stress.

Sinularia spp. *(S. asterolobata, S. flexibilis, S. polydactyla, etc.)*

Leather finger corals

PROFILE

The leather finger coral grouping contains several species with branching extensions of their tissue that give them their common name. They can form massive colonies, even in an aquarium, and need plenty of space to grow into.

WHAT size?

Wild specimens may cover several metres, but aquarium individuals usually remain smaller. However, where space allows they can form colonies measuring over 1m across, unless you control them by taking cuttings.

WHAT does it eat?

Primarily photosynthetic in the marine aquarium, but may also consume very fine particulate food.

WHERE is it from?

Tropical Indo-Pacific.

WHAT does it cost?

★★☆☆☆ ★★★★☆

Price depends on origin, size and colour.

SIMILAR SPECIES

Several species are sold as leather finger corals, but the husbandry and aquarium requirements for each are practically identical.

HOW compatible with other invertebrates?

Will not harm other invertebrates directly, but their prodigious growth rate can mean that corals located beneath the extended fingers are quickly starved of light.

WHAT water flow rate?

Moderate to high. Need the assistance of water flow to help them shed mucus. The intensity of the water currents directed towards them can determine their growth form. In strong currents, some species grow into quite squat forms with thick columns, whereas a more moderate flow results in the formation of highly branched, tall colonies.

HOW much light?

Moderate to strong illumination. Will thrive under strong T8, T5 or metal-halide lighting of sufficient intensity.

WILL fish pose a threat?

These corals secrete noxious defensive compounds, so very few fish will nibble at them.

WILL it threaten fish?

No, harmless to fish.

WHAT to watch out for?

Avoid corals showing patches

▲ *The rounded, fingerlike projections of this Sinularia sp. leather coral is only one of the many growth forms in this genus of leather corals.*

of discoloration or signs of powdery deposits around their base. When settled, these highly desirable corals contract their polyps in waves all over their surface, creating a wonderful, captivating display. However, some specimens may rarely 'pulse' their polyps.

WILL it reproduce in an aquarium?

Forms daughter colonies through budding of the main column or by shedding fingerlike extensions of tissue.

Sinularia dura

Cabbage coral

PROFILE

A hardy species of leather coral that forms encrusting colonies that can spread prolifically over suitable substrates. Colours available range from pink-brown and yellow to fluorescent green.

Cabbage coral is an attractive and hardy species and makes an excellent choice for beginners to the reefkeeping hobby. Try to avoid very pale specimens.

WHAT size?
Wild specimens may cover several metres. Aquarium individuals usually remain smaller, albeit often because the aquarist has controlled their growth by taking cuttings.

WHAT does it eat?
Primarily photosynthetic in the marine aquarium, but will also consume very fine particulate food.

WHERE is it from?
Tropical Indo-Pacific.

WHAT does it cost?
★★☆☆☆   ★★★★☆
Price depends on origin, size and colour.

SIMILAR SPECIES
The genus *Sinularia* is highly variable in appearance, but few members can be easily confused with *S. dura*. The most similar-looking corals are actually leather corals, such as *Lobophytum crassum*.

HOW compatible with other invertebrates?
The coral's steady growth may overshadow other invertebrates, but it does little direct harm even to its immediate neighbours.

WHAT water flow rate?
Moderate to high. Needs the assistance of water flow to help it shed mucus.

HOW much light?
Moderate to strong illumination. Will thrive under strong T8, T5 or metal-halide lighting of sufficient intensity. Less demanding of strong light than some other soft corals.

WILL fish pose a threat?
Few fish nibble at the flesh of this coral, although some will peck at expanded polyps. When this occurs, the bulk of the coral's tissue will remain expanded and show no other adverse reaction, other than to keep the affected polyps retracted.

WILL it threaten fish?
No. This coral is completely harmless.

WHAT to watch out for?
Avoid individuals showing yellow patches anywhere on the upper surface as these can indicate bacterial infections incurred through the rigours of shipping. Specimens may frustrate their owners by refusing to expand their polyps, despite appearing to be otherwise healthy and growing at this time. The reasons for this are unknown.

WILL it reproduce in an aquarium?
Cabbage coral will form new colonies that bud off from the main organism. It can be propagated by cuttings.

Studeriotes spp.
Christmas tree coral

PROFILE

A specialised coral that contains no symbiotic algae and therefore must be fed. There are white-brown and black forms. It usually lives in soft or rubble-based substrates, rather than on rock, and contracts into its base when disturbed. Its aquarium requirements are a challenge for any aquarist.

WHAT size?
Does not usually grow any taller than 20cm.

WHAT does it eat?
Relies entirely on food occurring naturally in the aquarium or added by the aquarist. Offer regular feedings of microplankton when the polyps are expanded. Be aware that this coral may only expand at certain times during the day or night.

WHERE is it from?
Tropical Indo-Pacific.

WHAT does it cost?
★★☆☆☆ ★★★☆☆
Inexpensive.

SIMILAR SPECIES
Not easily confused with other soft corals, especially when seen in the unusual colour forms.

HOW compatible with other invertebrates?
In any territorial disputes with its neighbours, it is likely to be the victim rather than the aggressor. The large amounts of food that it requires on a daily basis can compromise water quality for the other aquarium inhabitants.

WHAT water flow rate?
Strong indirect flow is best for most individuals.

HOW much light?
Indifferent to lighting or its intensity, but it can favour opening when all the lights are off in the aquarium.

WILL fish pose a threat?
Few fish will attempt to nip at this coral.

WILL it threaten fish?
No, harmless to fish.

WHAT to watch out for?
Ailing specimens look weedy and appear reluctant to expand their polyp tentacles. Recently imported specimens are probably the best ones to try, as they will not have been starved for as long as some displayed in the dealer's aquarium. Some dealers do offer fed specimens.

WILL it reproduce in an aquarium?
None recorded, possibly due to the poor rates of success experienced with this coral.

◁ *The round ball of tissue created when the coral is retracted needs to planted in the substrate to provide stability, as it has a high centre of gravity when the animal is expanded.*

Acrozoanthus spp.

Polyp tree

PROFILE

The polyp tree has a stunning appearance, is relatively cheap to buy and commonly available. However, the success rate for keeping it long term is low, perhaps due to a poor understanding of its aquarium requirements.

WHAT size?
Polyps measure around 10mm stalk diameter, but the tentacle disc can be over 8cm across in strong flow.

WHAT does it eat?
Requires plentiful feeding with fine particulate material. Any zooplankton substitutes will be accepted and result in good growth. The polyp tree contains photosynthetic algae.

WHERE is it from?
Tropical Indo-Pacific.

WHAT does it cost?
★★☆☆☆   ★★★☆☆
Price depends on the size of the colony.

SIMILAR SPECIES
Similar in overall appearance to the yellow polyp (see page 55). Colonies are available in green or brown morphs.

HOW compatible with other invertebrates?
Capable of stinging its neighbours, so allow space for the polyps to expand into.

WHAT water flow rate?
Good to strong indirect flow. The polyps will increase the length of their tentacles in stronger currents.

HOW much light?
Good to high. Will thrive under T5 or metal-halide lighting of sufficient intensity.

WILL fish pose a threat?
Some otherwise reef-compatible species may occasionally nibble at the tentacles of the polyps. Those that create most problems tend to be the species with a reputation for harassing certain corals, such as dwarf angelfish. However in most cases, the tree polyp will remain unmolested.

WILL it threaten fish?
May inflict a small amount of damage through their sting, but rarely threaten the life of any fish, however small.

WHAT to watch out for?
Imported specimens are polyps that have colonised the tubes of the polychaete worm *Eunice*. These degrade over time and polyps are lost as they no longer have any secure attachment site. However, when provided with

▲ *The beautiful polyps of* Acrozoanthus *need feeding in order to thrive and tend not to prosper in nutrient-poor systems.*

strong illumination and plentiful food, they will grow on almost any substrate.

WILL it reproduce in an aquarium?
Asexual reproduction is commonplace. Often polyps are shed from the main colony and these can re-attach and begin new growth and reproduction.

Anthelia spp.

Feather star polyp

PROFILE

This polyp forms creeping mats of small to large polyps, ranging from 20mm high to as long as 15cm, depending on the species. Under favourable conditions it spreads rapidly, but can sometimes fail to flourish despite being offered what – on paper – appear to be optimal conditions and positioning.

WHAT size?
Polyps will spread over almost any suitable hard surface, including glass, and sometimes even sand or gravel.

WHAT does it eat?
Traps very small particulate food from the water column, but appears to thrive in the absence of any supplemental feeding.

WHERE is it from?
Tropical Indo-Pacific.

WHAT does it cost?
★★☆☆☆ ★★★★☆
Price depends on size and colour variety. Feather star polyp is expensive, as it has large polyps and is difficult to ship. Give it time to settle before buying.

▶ Anthelia, *particularly the large-polyp species shown here, is a beautiful colonial polyp. In favourable conditions it can spread quickly onto a variety of substrates.*

HOW compatible with other invertebrates?
A mostly benign species. It may overgrow other corals, but this is rare.

WHAT water flow rate?
Moderate to high. *Anthelia* spp. enjoy intermittent currents, but these are not necessary for their long-term survival.

HOW much light?
Moderate to high. Will thrive under T5 or metal-halide lighting.

WILL fish pose a threat?
Some dwarf angelfish have been observed pecking at newly introduced colonies, but most will leave the polyps alone.

WILL it threaten fish?
No. This polyp is not known to harm any fish species.

WHAT to watch out for?
Watch out for colonies where a number of the polyps present have retracted tentacles, as they are unlikely to have recovered fully from the rigours of shipping.

WILL it reproduce in an aquarium?
Asexual reproduction is very common. Polyps will bud from the colony or detach completely in order to exploit new, as yet uncolonised, substrates.

SIMILAR SPECIES

Most similar to feather star polyps are members of the genus *Clavularia*, the clove, jasmine or daisy polyps (see page 50).

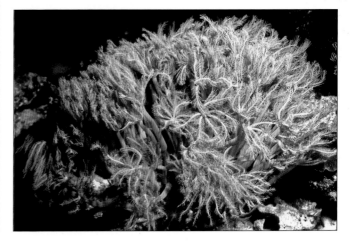

Clavularia spp.

Clove star polyp

PROFILE

A variable genus in which the colours range from creamy brown to brilliant green; the latter is rare and highly sought-after. The polyps retract into small egg-shaped sheaths at night or when disturbed, thereby differentiating them from similar species, such as *Anthelia* spp.

WHAT size?
Polyps will spread over almost any suitable hard surface, including glass and sometimes even sand or gravel. Polyp size depends on the species concerned. Some seldom reach more than 1cm, whereas others may achieve 10cm or more.

WHAT does it eat?
Traps very small particulate food from the water column, but appears to thrive in the absence of any supplemental feeding.

WHERE is it from?
Tropical Indo-Pacific.

WHAT does it cost?
★★☆☆☆ ★★★★☆
Price depends on the size of the colony and the colour.

▶ *Although clove star polyps resemble* Anthelia *spp. when expanded, they retract into a sheath, which* Anthelia *does not.*

HOW compatible with other invertebrates?
Its growth rate means that it could threaten some slow-growing stony corals, but this is rare.

WHAT water flow rate?
Clavularia enjoy moderate to strong currents.

HOW much light?
Moderate to high. Will thrive under T5 or metal-halide lighting.

WILL fish pose a threat?
Some fish may peck at the tentacles, but most species deemed to be reef-safe will ignore it entirely.

WILL it threaten fish?
No. This polyp is not known to harm any fish species.

WHAT to watch out for?
Colonies should have few, if any retracted polyps, and the tentacles should be well spread out.

WILL it reproduce in an aquarium?
Reproduces sexually almost constantly, provided it has suitable conditions, including bare substrate, to colonise.

SIMILAR SPECIES

Most similar to clove star polyps are members of the genus *Anthelia*, the feather star polyps. Pulse corals (*Xenia/Heteroxenia* spp.) are also similar (see page 59).

Discosoma spp.

Mushroom polyp

PROFILE

Members of the genus *Discosoma* are available in a wide variety of colour morphs. They are tolerant of a wide range of light intensities and are generally undemanding.

WHAT size?
Large specimens reach 10cm or so when fully stretched.

WHAT does it eat?
Traps food particles from the water column and ingests them through the central mouth, but also houses photosynthetic symbionts.

WHERE is it from?
Tropical Indo-Pacific.

WHAT does it cost?
★★☆☆☆ ★★★★★
Price depends on size and colour variety.

SIMILAR SPECIES

Mushroom polyps from the genus *Discosoma* are available in a wide variety of colours and many have different textures to the surface of the animal. Similar species include *Rhodactis*, *Ricordea* and *Amplexidiscus*.

▲ *The blue and brown specimens shown here are* Discosoma *spp. and the green specimens to the right are* Rhodactis *spp. (See page 56.)*

HOW compatible with other invertebrates?
A benign invertebrate that tolerates its cohabitants and seems to be accepted by them. Placing a specimen downstream from a soft coral that secretes noxious defensive chemicals can upset them, and means they might evacuate from their rock.

WHAT water flow rate?
Low to moderate. Will often not expand fully when subjected to high flow rates.

HOW much light?
Low to moderate. Thrives towards the bottom of most aquariums.

WILL fish pose a threat?
Most reef-compatible fish will not bother mushroom polyps. Dwarf angelfish might occasionally nip at a specimen, but this is unusual.

WILL it threaten fish?
No. Even large specimens will not attempt to predate fish.

WHAT to watch out for?
Several importers offer propagated colonies for sale, sometimes with wonderful colour variations. Combining different varieties of this and other mushroom polyp species makes for a wonderful display.

WILL it reproduce in an aquarium?
Yes. Reproduction is commonly asexual through budding or splitting of a single polyp. Sexual reproduction is rarer but has been reported in captivity.

Isaurus sp.

Snake polyp

PROFILE

An interesting genus, characterised by the nocturnal opening of the polyps. They are able to gain energy through photosynthesis during the day and must have strong illumination. Given time and good husbandry, the polyps may open during the day, particularly if they are fed at this time.

WHAT size?
Polyps will spread over almost any suitable hard surface, including rock and glass and sometimes even sand or gravel. Each polyp seldom grows more than 10cm long.

WHAT does it eat?
Primarily zooplankton, but will gain most of its nourishment from its photosynthetic algae. If the polyps only open at night, then feed 2-3 times per week on cyclops, mysis or brineshrimp. If and when they begin to open during the day, feed them at the same time as the other livestock.

WHERE is it from?
Circumtropical. Most aquarium specimens are collected from the Indo-Pacific region.

WHAT does it cost?
★★☆☆☆   ★★★☆☆
Price depends on size.

HOW compatible with other invertebrates?
Should not affect other invertebrates, although probably able to hold its own against other polyps as they grow and begin to compete with the snake polyps for available substrate.

WHAT water flow rate?
Moderate to high. Strong flow can make feeding the polyps difficult.

HOW much light?
Good to high. Will thrive under T5 or metal-halide lighting.

WILL fish pose a threat?
Most reef-compatible fish will ignore the snake polyp, but some may find an open polyp too tempting not to nibble.

WILL it threaten fish?
No. This polyp is not known to harm any fish species.

WHAT to watch out for?
You are unlikely to observe specimens with open polyps in a dealer's tank. Therefore, check

▲ *Initially, most aquarists will encounter the snake polyp as it can be seen here, with the polyp tentacle crown retracted.*

the knobbly stalks for signs of damage or discoloration. If these are absent, you can buy the polyp with a fair degree of confidence.

WILL it reproduce in an aquarium?
Asexual reproduction is very common. Polyps will bud from the colony or detach completely in order to exploit new, as yet uncolonised, substrates.

SIMILAR SPECIES
Although not unique, this genus is unlikely to be confused with any other that is commonly available in the aquarium hobby. Expanded specimens can resemble colonies of *Protopalythoa*.

Pachyclavularia viridis

White spot star polyp

PROFILE

Ranging from deep brown through green to a pale pink that appears almost silver, white spot star polyp is a firm favourite with reef aquarists. It is capable of prodigious growth and the tentacles can elongate, giving a real impression of movement in the aquarium. Metallic green forms are highly desirable.

WHAT size?
Individual polyps may measure less than 1cm across the tentacles and 1.5cm tall, but colonies can form creeping mats that spread almost anywhere in the aquarium, as long as there is substrate to colonise. This includes rockwork, glass and electrical hardware!

WHAT does it eat?
Traps very small particulate food from the water column, but appears to thrive in the absence of any supplemental feeding.

WHERE is it from?
Tropical Indo-Pacific.

WHAT does it cost?
★★☆☆☆   ★★★★☆
Price depends on size and colour variety.

HOW compatible with other invertebrates?
Be sure to protect any coral that could be overgrown by this polyp. It does not have a strong sting, but can overwhelm others simply by growing more quickly than they can.

WHAT water flow rate?
Prefers moderate to high flow. Tolerates low flow, but does not like to experience sedimentation on its surface, so take steps to avoid this.

HOW much light?
Moderate to high. Will thrive under T8, T5 or metal-halide lighting. The latter may be necessary to maintain the

SIMILAR SPECIES
There are several species known as 'star polyps', including *Briareum* spp. and many others from the genus *Pachyclavularia*. Those that form encrusting mats are very similar in terms of their husbandry.

colours of metallic green specimens.

WILL fish pose a threat?
Not many fish will nip at star polyps, although there will be rogues of any species that take a liking to it.

WILL it threaten fish?
This polyp is not known to harm any fish species.

WHAT to watch out for?
Colonies should have all the polyps out and expanded. Avoid polyps that appear to have lost colour or are very small.

WILL it reproduce in an aquarium?
Asexual reproduction is normal and widespread. Has been reported spawning frequently.

◄ *Metallic green specimens are always popular. They spread onto almost any hard substrate, including the aquarium glass.*

Palythoa and *Protopalythoa* spp.

Giant button polyp

PROFILE

A large and commonly kept species of colonial polyp, available in brown and green. The metallic green forms are particularly desirable; when the polyp tentacles are extended they resemble Venus fly traps.

WHAT size?
Individual polyps can measure up to 2.5cm across, including their tentacles, and about 1.5cm tall. Colonies can grow as large as space allows in the aquarium.

WHAT does it eat?
Contains photosynthetic pigments, but will also consume particulate food up to and including the size of a mysis shrimp or similar.

WHERE is it from?
Tropical Indo-Pacific.

WHAT does it cost?
★★☆☆☆ ★★★★☆
Price depends on the size of the rock that plays host to the colony, and the colour. Metallic green specimens command the highest prices.

◀ *In settled specimens, the tentacles surrounding the margin of the polyp will often expand significantly, making for an extremely attractive appearance.*

HOW compatible with other invertebrates?
Where conditions allow, it can be overgrown by fast-growing invertebrates. Colonial tunicates and sponges can threaten them where present.

WHAT water flow rate?
Moderate to high. This polyp does not like to be placed in areas with little or no current, where it can become fouled by detritus and algae.

HOW much light?
Moderate to high. Will thrive under T8, T5 or metal-halide

lighting that provides sufficient light intensity.

WILL fish pose a threat?
Tangs, surgeonfish, angelfish and dwarf angelfish have all been observed nipping at the tentacles of giant button polyps. This often causes the tentacles to curl in around their outer margin.

WILL it threaten fish?
Not known to be harmful to any fish species.

WHAT to watch out for?
Avoid specimens with closed polyps. The tentacles should be stretched and a solid colour. Ignore those showing white patches in the centre of any of the polyps.

WILL it reproduce in an aquarium?
Asexual reproduction is common and the colony will steadily grow into the surrounding area. Will colonise bare rock and glass with ease.

SIMILAR SPECIES
The largest 'giant polyp' is actually *Palythoa grandis*, which can easily reach a diameter of over 2.5cm. Husbandry is the same as for the species listed here. They also resemble several species of zoanthid polyp.

Parazoanthus gracilis

Yellow polyp

PROFILE

It is best to think of yellow polyps as small colonial anemones. This beautiful creature will thrive in the home aquarium provided it receives sufficient light, as well as regular feeding.

WHAT size?

Individual polyps usually measure less than 2cm across the tentacles and 2cm tall, but given the opportunity, colonies can extend over vast areas. They will spread onto glass or almost any exposed surface.

WHAT does it eat?

Traps small particulate material suspended in the water column. Benefits from occasional feedings with brineshrimp, their nauplii and proprietary invertebrate products such as frozen marine copepods.

WHERE is it from?

Tropical Indo-Pacific.

WHAT does it cost?

★☆☆☆☆   ★★★☆☆
Price depends on the size of the rock that plays host to the colony.

▶ *Yellow polyps are an inexpensive and attractive addition to the reef aquarium, but rapid expansion of the colony can lead to problems with its neighbours if allowed to proceed unchecked.*

SIMILAR SPECIES

Acrozoanthus, the 'tree polyp', is a brown colonial polyp similar in overall appearance to the yellow polyp. It is more difficult to maintain in the home aquarium, although aquarists offering regular feedings of zooplankton substitutes and phytoplankton report great success with this animal (see page 48).

HOW compatible with other invertebrates?

Yellow polyps can sting their neighbours, so it is worth leaving a clear area around the original colony to allow them to spread.

WHAT water flow rate?

Moderate to high. This polyp does not like to be placed in areas with little or no current.

HOW much light?

Moderate to high. Will thrive under T8, T5 or metal halide lighting that provides sufficient light intensity.

WILL fish pose a threat?

A few fish will nip these polyps, particularly those that have been stocked to 'control' the pest anemone *Aiptasia* sp.

WILL it threaten fish?

May inflict a small amount of damage through their sting, but rarely threaten the life of any fish, however small.

WHAT to watch out for?

Avoid specimens with a number of the polyps closed or colonies that are not a vibrant yellow. In healthy specimens, polyp tentacles should be expanded and stretched.

WILL it reproduce in an aquarium?

Asexual reproduction is commonplace.

Rhodactis spp.

Mushroom anemone

PROFILE

A highly variable genus, with tentacles ranging from very short to substantially long in some species. There is a wide range of colour forms from brown to green, plus the rarer blue and purple forms.

WHAT size?
Large specimens reach 10-15cm or so when fully expanded.

WHAT does it eat?
Feeds anemone-like on particulate material. Can be target-fed, but often thrives without specific intervention on the part of the aquarist.

WHERE is it from?
Tropical Indo-Pacific.

WHAT does it cost?
★☆☆☆☆   ★★★★☆
The price range given covers small pieces or rocks with only one or two polyps present up to larger specimens.

SIMILAR SPECIES
Almost any mushroom polyp could be confused with *Rhodactis*, including *Ricordea* spp. *Discosoma* spp. and *Amplexidiscus* sp.

▲ Rhodactis *spp. can be distinguished from* Discosoma *spp. by the presence of tiny tentacles over the surface of the polyp. Sometimes these are grouped in clusters.*

HOW compatible with other invertebrates?
Capable of stinging its neighbours but will usually coexist peacefully with most. Those with longer tentacles present the greatest threat to their neighbours, but compared with other, more aggressive corals, their impact is fairly low.

WHAT water flow rate?
Moderate to good. The longer the tentacles, the more tolerant of high flow the individual will be.

HOW much light?
Low to medium.

WILL fish pose a threat?
Some fish find the longer-tentacled species just too tempting to resist and will repeatedly peck at the trailing tissue.

WILL it threaten fish?
May predate very small fish (less than 2cm) but this is very rare indeed.

WHAT to watch out for?
Unhealthy individuals often display patches of white tissue or contracted tentacles. Allow plenty of room for growth as they can reproduce quickly.

WILL it reproduce in an aquarium?
Yes. Asexual splitting or budding is commonplace and spawning events have been recorded.

Ricordea spp.

Knobbly mushroom

PROFILE

This species has a number of short, rounded tentacles across the surface of the disc. Most Indo-Pacific specimens are sold as small colonies, but those collected from the Caribbean and surrounding areas are available singly. This is due to restrictions that forbid the export of 'live' rock from the tropical Atlantic.

WHAT size?
Large specimens reach 10-15cm or so when fully expanded.

WHAT does it eat?
Feeds anemone-like on particulate material. Seems to thrive with or without target-feeding.

WHERE is it from?
Tropical Indo-Pacific, Tropical Atlantic, Caribbean

WHAT does it cost?
★☆☆☆☆   ★★★☆☆
Price range given is for Indo-Pacific colonies, depending on their size, density of polyps and coloration. Individual polyps vary in price, depending on their size and colour. Bright green, red and turquoise are among the most desirable forms.

HOW compatible with other invertebrates?
Capable of stinging its neighbours, but is usually under more threat of being stung by others.

WHAT water flow rate?
Low to moderate.

HOW much light?
Low to medium.

WILL fish pose a threat?
Dwarf angelfish may nip at them, but most reef-compatible species will ignore them entirely. Anemonefish occasionally seek refuge in colonies in the absence of a suitable host anemone.

WILL it threaten fish?
May predate very small fish (less than 2cm), but this is very rare indeed.

SIMILAR SPECIES
Almost any mushroom polyp could be confused with *Ricordea*, including *Rhodactis* spp. *Discosoma* spp. and *Amplexidiscus* sp.

WHAT to watch out for?
Avoid specimens that appear contracted or those with areas of clear tissue (a sign of pigment loss).

WILL it reproduce in an aquarium?
Yes. Asexual splitting or budding is commonplace and spawning events have been recorded, despite being rare.

▼ *Ricordea* spp. *are becoming increasingly popular with aquarists. Green, red and blue specimens are always highly sought after.*

Tubipora musica

Organ pipe coral

PROFILE

Although this coral secretes a stony skeleton of sorts, it is more closely related to colonial polyps and soft corals than to the true stony corals. Generally, individuals with larger polyps tend to be easier to maintain in an aquarium, but this can be a temperamental species that confounds the most diligent aquarists.

WHAT size?
Wild individuals may reach over 1m across, but aquarium specimens seldom reach these proportions.

WHAT does it eat?
Primarily photosynthetic in the aquarium, but will also take very fine particulate food.

WHERE is it from?
Tropical Indo-Pacific.

WHAT does it cost?
★★☆☆☆   ★★★★☆
Price depends on size and colour.

HOW compatible with other invertebrates?
Does not present any threat to other corals or invertebrates.

WHAT water flow rate?
Strong indirect flow is essential for the long-term well-being of this coral

HOW much light?
Demands strong illumination. T5 is an absolute minimum, but most specimens need metal-halide lighting of higher wattages.

WILL fish pose a threat?
Most fish will ignore this coral. It can be upset by fish browsing on some of the incidental organisms present on its hard skeleton, which causes the polyps to remain retracted.

WILL it threaten fish?
No. This coral is completely harmless.

WHAT to watch out for?
This coral hates to become smothered with detritus or

colonised by algae so a nutrient-poor aquarium with good mechanical filtration is a must. It favours conditions often reserved for stony corals.

WILL it reproduce in an aquarium?
Has the potential to spawn in the aquarium, but this is rarely reported. Most reproduction is asexual, but given the species' poor record for survival in the home aquarium even this is rarely achieved.

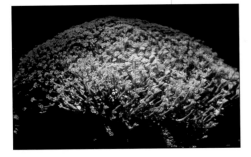

◀ The hard skeleton is clearly visible here and gives this animal its common name of organ pipe coral.

▶ Polyp form is highly variable in organ pipe corals.

Xenia and *Heteroxenia* spp.

Pulse corals

PROFILE

Ranging from almost white to rich brown, the pulse corals are aquarium favourites. As well as their obvious beauty, they 'pulse', contracting their tentacle crown at regular intervals using internal musculature. However, their ability to colonise every substrate means they must be harvested regularly if they are not to overrun the aquarium.

WHAT size?

Polyps will spread over almost any suitable hard surface, including glass and sometimes even sand or gravel. Polyp size depends on the species concerned. Some have long stalks measuring several centimetres, whereas others form lower, encrusting mats.

WHAT does it eat?

Traps very small particulate food from the water column, but appears to thrive in the absence of any supplemental feeding.

WHERE is it from?

Tropical Indo-Pacific.

WHAT does it cost?

★★☆☆☆   ★★★★☆

Price depends on the size of the colony and its origin. Cultured specimens are often very cheap, as aquarists literally give them away when they overgrow their aquarium.

HOW compatible with other invertebrates?

Its growth rate means that it could threaten some slow-growing stony corals, but this is rare. Check wild-collected colonies for signs of hitch-hiking predators, such as crabs and polychaete worms.

WHAT water flow rate?

Moderate to high. *Xenia,* particularly, enjoy intermittent currents. A period of slack water will often encourage them to pulse vigorously.

HOW much light?

Good to high. Will thrive under T5 or metal-halide lighting. Aquarium-cultured specimens are widely available and often prove more tolerant of lower light levels than their wild-collected cousins.

WILL fish pose a threat?

Some fish may peck at the

▲ *This* Xenia *spp. only has a single polyp form, whereas* Heteroxenia *spp. has two. Pulse corals are often cultured in home aquariums due to their prolific growth rate.*

tentacles, but most species deemed to be reef-safe will ignore it entirely.

WILL it threaten fish?

No. Pulse corals are not known to harm fish, even though they have high concentrations of iodine in their tissues, which is thought to deter predation.

WHAT to watch out for?

Tentacles should be stretched and show no signs of discoloration. Some species are particularly prone to bacterial infections and do not ship well, so give them plenty of time to settle before buying them.

WILL it reproduce in an aquarium?

In suitable conditions, including bare substrate to colonise, it reproduces sexually almost constantly.

SIMILAR SPECIES

Xenia and *Heteroxenia* as often confused with *Cespitularia*, or African blue coral, as it is sometimes known. The latter is often more difficult to maintain, demanding good flow and strong illumination together with excellent water quality.

Zoanthus spp.

Button polyps, or sea mats

These colonial polyps are some of the hardiest and most commonly kept marine invertebrates. They are available in a wide range of colour varieties from brown to green and pink to orange. Keeping several in close proximity creates a wonderful display.

WHAT size?
Individual polyps can measure up to 1.5cm across, including their tentacles and about 1cm tall. Colonies can grow as large as space allows in the aquarium.

WHAT does it eat?
Contains photosynthetic pigments, but also consumes particulate food, usually no larger than brineshrimp.

WHERE is it from?
Tropical Indo-Pacific.

WHAT does it cost?
★★☆☆☆ ★★★★☆
Price depends on size and colour.

SIMILAR SPECIES
Resembles members of the genera *Palythoa* and *Protopalythoa,* also known as button polyps.

HOW compatible with other invertebrates?
Susceptible to stinging attacks by stony corals, but will coexist with most stony corals and other colonial polyps, even when they are immediately adjacent.

WHAT water flow rate?
Moderate to high. This polyp does not like to be placed in areas with little or no current, where it can become fouled by detritus and algae.

HOW much light?
Moderate to high. Will thrive under T8, T5 or metal-halide lighting. In systems with more intense lighting, they can be placed towards the bottom of the aquarium.

Connoisseurs of the humble button polyp will seek out unusual colour varieties, such as red or blue, as well as those containing fluorescent pigments. Several are shown here.

WILL fish pose a threat?

Some surgeonfish have been observed attacking polyps, but such experiences are rare. Most reef-compatible fish find this polyp unpalatable.

WILL it threaten fish?

Will not harm any fish species.

WHAT to watch out for?

Avoid specimens with closed polyps or those that appear pale. Tentacles should be expanded on all specimens.

WILL it reproduce in an aquarium?

Asexual reproduction is commonplace and the colony will steadily grow into the surrounding area. Will colonise bare rock and glass with ease.

Founders of the reef

Stony corals are so-named because they are able to secrete a calcium carbonate exoskeleton. Of the 3,500 species currently described, the majority contain zooxanthellae from which the animal derives a significant proportion of its nutrition. However, recent studies indicate that stony corals on natural reefs will also capture significant quantities of planktonic food from the water column. In general terms, each group requires slightly different conditions. Small-polyp stony corals (SPS) tend to prefer stronger illumination and water currents than large-polyp stony corals (LPS), but in most systems suitable for SPS there will be areas where LPS can be easily accommodated. There is a tendency in the hobby to regard LPS as 'easier' than SPS, but both require good, stable water conditions and diligent maintenance if they are to thrive.

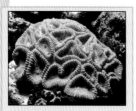

Price guide

★	£10 – 20
★★	£20 – 40
★★★	£40 – 60
★★★★	£60 – 80
★★★★★	£80 – 100

PROFILE

A diverse genus of stony corals with a wide range of growth forms, from low and stumpy to tabular, highly branching and bushy. Some species show a variety of growth patterns depending on where they are situated, for example in high or low flow.

WHAT size?

Massive colonies can form on natural reefs and their growth is similarly boundless in a home aquarium. Most people take cuttings of their corals before they grow too large.

WHAT does it eat?

In most aquariums, the coral depends on strong illumination to stimulate the production of sugars by its zooxanthellae. Wild colonies also take significant amounts of food from the water column. Microplankton substitutes added regularly are extremely useful in maintaining healthy colonies of *Acropora*.

WHERE is it from?

Circumtropical.

WHAT does it cost?

★☆☆☆☆   ★★★★★
Acropora is more widely available and reasonably priced than it was ten years ago. Pieces range from frags to wild-collected colonies.

Acropora spp.

Stagshorn coral

Maintaining the beautiful colours of Acropora spp. *often requires care.*

HOW compatible with other invertebrates?

Not especially aggressive, but can defend itself against similar species. Sometimes uses its digestive tissues to eat away at the flesh of a neighbour that has grown too close to it.

WHAT water flow rate?

Most *Acropora* species prefer strong currents.

HOW much light?

T5 lamps or metal-halide lights are essential. Depending on the wattage and the species involved, the coral may have to be placed towards the top of the aquarium to receive sufficient light. Light influences the health of the coral and its coloration. Many corals utilise protective pigments that fluoresce, reflecting back harmful UV radiation from sunlight.

WILL fish pose a threat?

Some fish deemed to be generally reef-safe may occasionally peck at *Acropora*. The attractive coral gobies (*Gobiodon* spp.) have evolved to live amongst *Acropora* corals. Some species feed on zooplankton, but many include the mucus or flesh of their host in their natural diet.

WILL it threaten fish?

This coral should not harm fish.

WHAT to watch out for?

Ignore excessively pale corals or those with exposed white skeleton. Sometimes, pale colonies are entirely natural, being found in shallow water. However, they also indicate stressed corals that have jettisoned their zooxanthellae.

WILL it reproduce in an aquarium?

Yes, but rarely. There seems to be a minimum size at which certain species mature, but many aquarium specimens will not achieve this. Many corals and other invertebrates begin spawning as a result of external stimuli, such as the lunar cycle, but in effect these are absent from most home aquariums.

▼ *Cuttings of* Acropora *and similar small-polyp stony corals are called 'frags' and are often taken by aquarists when pruning a specimen.*

MANY SPECIES

The number of *Acropora* species currently described is well over 100 and rising all the time. All require largely the same care, but for wild-collected specimens, pigmentation and growth form provide clues to their correct positioning in terms of flow and light intensity.

Alveopora spp.

Alveopora, or flowerpot, coral

PROFILE

Alveopora is stunning to behold, particularly the metallic green individuals imported sporadically. It has 12 short, stout tentacles arising from a fairly fragile skeleton. The polyps of *Goniopora*, its close relative, have 24 tentacles.

WHAT size?
Maximum growth in the aquarium is uncertain as few aquarists manage to maintain this animal for prolonged periods, but in successful cases it has the potential to grow over 30cm in diameter.

WHAT does it eat?
Derives its primary nourishment from symbiotic algae contained in its tissue. Many aquarists believe that feeding also plays a part in success with this animal, advocating regular additions of phytoplankton and very small zooplankton.

WHERE is it from?
Tropical Indo-Pacific.

WHAT does it cost?
★★☆☆☆   ★★★☆☆
Price depends on size and colour. At present, most specimens are collected from the wild or cultured abroad, for example in Fiji or Indonesia.

HOW compatible with other invertebrates?
The polyps can inflict a powerful sting, but the tissue of this animal is very thin and does not fare well once damaged, either through poor handling or as a result of stinging attacks from its neighbours. Allow plenty of room for polyp expansion and, hopefully, growth.

WHAT water flow rate?
Moderate flow. If the polyps do not extend much, the coral is probably in too much current.

HOW much light?
Moderate to strong illumination, preferably under metal-halide or multiple T5 lamps.

WILL fish pose a threat?
In the absence of a suitable host anemone, some anemonefish may single out the coral. This can irritate the coral and prevent it from expanding its polyps.

WILL it threaten fish?
Capable of stinging careless fish, but such events are rare and unlikely to lead to any lasting problems for either the coral or its victim.

▶ *Resembling a vase full of daisies, it is easy to see how the flowerpot coral acquired its common name.*

WHAT to watch out for?
It is best to seek out colonies with expanded polyps and no sign of any pure white skeleton. Polyps can retract quite far into the skeleton, leaving little evidence of living tissue. This does not necessarily indicate an unhealthy specimen, but neither does it identify a healthy one!

WILL it reproduce in an aquarium?
Has the potential to drop fully formed colonies containing a small number of polyps and a tiny piece of skeletal material.

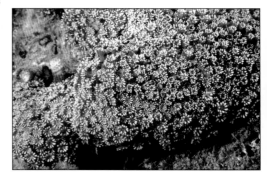

Caulastrea spp.

Trumpet coral

PROFILE

Large fleshy polyps at the end of quite long branches typify this coral. It is widely available and very popular amongst aquarists on account of its hardy disposition.

WHAT size?

Colonies usually grow to about 30cm in diameter, but aquarists often take cuttings before they reach this size.

WHAT does it eat?

Contains symbiotic algae that satisfy the majority of its nutritional requirements. The presence of suitable foods in the water will encourage the feeding tentacles to come out to try to capture this bounty. Frozen marine copepods and other plankton feeds can achieve this. Once the tentacles are everted, you can place larger food particles, such as mysis and brineshrimp, directly onto them.

WHERE is it from?

Tropical Indo-Pacific.

WHAT does it cost?

★★☆☆☆

Price depends on size and colour, as well as the point of origin. Brown and green specimens are the least expensive, but cultured 'super green' fluorescent colonies can be costly.

▲ *Caulastrea* spp. are not particularly variable in form or colour, but those containing fluorescent pigments often command a higher price.

HOW compatible with other invertebrates?

Can use its tentacles in defence or attack, but is generally not sufficiently well equipped to withstand aggression from other species. Allow space around the animal, not only for growth and polyp expansion, but also as a protective buffer zone.

WHAT water flow rate?

Low to moderate. Can tolerate strong currents, but these may prevent full polyp expansion.

HOW much light?

Will thrive and grow under T5 and metal-halide lighting, although given sufficient wattage, it is possible to maintain this coral under T8 lamps. Specimens placed in strong light often develop white bands across the polyp, but this does not appear to threaten the health of the animal.

WILL fish pose a threat?

Some fish occasionally nibble at the coral's fleshy polyps. Dwarf angelfish and butterflyfish may be responsible, but it depends on the individuals concerned.

WILL it threaten fish?

No. When the tentacles are exposed, for example during feeding, it is capable of inflicting a sting, but there are no reports of fish being damaged.

WHAT to watch out for?

Polyps should be well expanded and none of the corallites (skeleton) should be visible. Occasionally, polyps are lost from wild specimens but the resulting exposed calcium carbonate is rapidly colonised by other animals. Bare white skeleton thus suggests that the die-off has occurred recently.

WILL it reproduce in an aquarium?

Spawning events are rare but have been reported for captive specimens.

SIMILAR SPECIES

Could be confused with *Blastomussa* spp., which also have short corallites, but that coral's polyps have distinctive oval-shaped swellings that are absent in *Caulastrea*. Can resemble *Favia* spp. when all the polyps are fully expanded so that no gaps are visible between them.

Euphyllia ancora

Hammerhead, or anchor, coral

The characteristic shape of the tentacles of this highly popular *Euphyllia* species accounts for the coral's common name. Some forms secrete a single massive skeleton, whereas others are branched. The former can be more difficult to place in the reef aquarium, especially when large.

WHAT size?
Grows rapidly in the wild as well as in the aquarium, where it can easily reach well over 60cm across the polyps. On natural reefs it may grow to more than one metre.

WHAT does it eat?
Primarily photosynthetic in the marine aquarium, but will also consume a wide variety of other items, from fine, particulate material to larger chunks of shellfish. Drop these with care directly into the centre of the tentacle mass.

WHERE is it from?
Tropical Indo-Pacific.

WHAT does it cost?
★☆☆☆☆   ★★★★☆
Small cultured specimens can be acquired quite cheaply, but larger, wild-collected colonies can be expensive.

HOW compatible with other invertebrates?
Euphyllia spp. have specialised tentacles with higher-than-usual concentrations of stinging cells. These 'sweeper tentacles' extend beyond the colony and can sting neighbouring animals, although not usually other euphyllia. Allow plenty of room for the almost inevitable growth of this animal in favourable conditions.

WHAT water flow rate?
Moderate to high. Avoid high-velocity direct currents, which can damage its delicate tissues or prevent polyps expanding.

HOW much light?
Will thrive under T5, metal-halide and T8 illumination of sufficient wattage.

WILL fish pose a threat?
Most reef-safe fish will ignore it entirely. An anemonefish can apparently irritate the animal if it adopts the coral as a surrogate anemone, thereby causing the tentacles to remain retracted.

WILL it threaten fish?
Is capable of stinging fish, particularly slow-moving species, but this seldom proves fatal or causes the victim long-lasting problems.

WHAT to watch out for?
Newly imported or poorly acclimatised specimens are prone to bacterial and protozoan

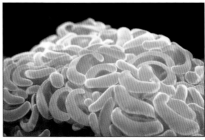

▲ *The hammerhead's unmistakable and beautiful polyps make this coral one of the most popular of all stony species.*

infections, often manifested by the presence of brown jelly around the living tissue. In good specimens the tentacles are expanded and only the lower portion of the stony skeleton is visible.

WILL it reproduce in an aquarium?
Will occasionally spawn, but most reproduction is by the dropping of polyp clusters that arise from the marginal tissue of the animal. It is very easy to take frags from branching colonies.

SIMILAR SPECIES
Several species of *Euphyllia* are commonly imported for the aquarium hobby. Of these, honey euphyllia *(Euphyllia sp.)* most closely resembles hammerhead coral, but can prove slightly less consistent than its more familiar relatives.

Euphyllia divisa

Frogspawn coral

PROFILE

A stunning coral, valued for the wonderful polyp structure that gives this animal its common name. It forms branching colonies, and ranges in polyp colour from creamy brown to brilliant green. It is typically less commonly available than many other *Euphyllia* species and therefore highly prized.

WHAT size?

Grows rapidly in the wild as well as in the aquarium, where it can easily reach well over 60cm across the polyps. On natural reefs it may grow to more than one metre.

WHAT does it eat?

Primarily photosynthetic in the marine aquarium, but will also consume a wide variety of other items, from fine, particulate material to larger chunks of shellfish. Drop these with care directly into the centre of the tentacle mass.

WHERE is it from?

Tropical Indo-Pacific.

WHAT does it cost?

★★☆☆☆   ★★★★☆

Small, cultured specimens can be acquired fairly cheaply, as aquarists with successful colonies are quite anxious to reduce the size of their specimen. Larger, wild-collected colonies can be expensive.

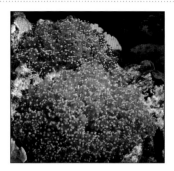

▲ *Take regular cuttings to prevent conflicts with E. divisa's neighbours.*

HOW compatible with other invertebrates?

Euphyllia spp. have specialised tentacles with higher-than-usual concentrations of stinging cells. These 'sweeper tentacles' extend beyond the colony and can sting neighbouring animals Allow plenty of room for growth.

WHAT water flow rate?

Moderate to high. Avoid high-velocity direct currents, which can damage the delicate tissues or prevent polyps expanding.

HOW much light?

Will thrive under T5, metal-halide and T8 illumination of sufficient wattage.

WILL fish pose a threat?

Few fish tackle this coral, especially as it has quite a potent sting. Even aquarists must take care. It can be irritated when anemonefish adopt it as a surrogate for a host anemone.

WILL it threaten fish?

Capable of stinging fish, particularly slow-moving species, but this is seldom fatal or causes long-lasting problems.

WHAT to watch out for?

Until its state of health is clear, avoid any specimen in which the polyps are not visible or uniformly expanded, or with signs of brown jelly anywhere on the living tissue. Signs of brilliant white heads where tentacles are absent might signify either an ailing specimen of one that has successfully recovered from disease.

WILL it reproduce in an aquarium?

Spawning is rare. Most colonies grow miniature facsimiles of the parent beneath the tentacle crown, where the wafer-thin tissue covers the calcium carbonate skeleton. These will often drop off by themselves, but with care, they can be removed manually and placed in an area where their subsequent growth and development can be monitored more easily.

SIMILAR SPECIES

Torch coral (*E. glabrescens*) is commonly available and very similar to keep. Its tentacles are a fairly uniform tubular shape, ending in a tip that can be white, cream, pink or green.

Favia spp.

Pineapple, or moon, coral

PROFILE

A beautiful, large-polyp stony coral, available in a variety of colour forms. In common with many such corals, it does not always extend the polyp tentacles. Instead, it keeps them in reserve for feeding or defence and/or attack.

WHAT size?

This genus can grow massive, with reports of colonies measuring several metres in diameter. However, as it grows slowly, most specimens acquired at 6-12cm should not outgrow the aquarium.

WHAT does it eat?

These corals are primarily photosynthetic and must be tempted to evert their tentacles. Often, sprinkling some food (say, brineshrimp or frozen marine copepods) directly onto them acts as an incentive. After 10-15 minutes the tentacles should be visible and you can offer more food.

WHERE is it from?

Tropical Indo-Pacific.

WHAT does it cost?

★☆☆☆☆ ★★★☆☆

For a stony coral this is an inexpensive species. Price depends on size and colour.

▶ Favia *spp. can be found in a wide variety of colour morphs. Those containing red or other fluorescent pigments are highly sought after.*

HOW compatible with other invertebrates?

Not particularly aggressive but will make its intentions known by everting its tentacles when it feels threatened. Some of these stretch to a remarkable length, so give them plenty of room.

WHAT water flow rate?

Moderate currents are sufficient.

HOW much light?

Best maintained under T5 or metal-halide lighting.

WILL fish pose a threat?

Some fish will peck at the coral's polyps. Dwarf angelfish are sometimes implicated, but it will depend on the individual and species concerned.

WILL it threaten fish?

No.

WHAT to watch out for?

Avoid bleached specimens or those with tissue missing from polyps. Die-off can be a natural

process, but bare white skeleton can indicate that it has happened recently, say, since importation.

WILL it reproduce in an aquarium?

Spawning events are reported for this coral. Although these are rare, it suggests that they may be more common for this group than for most other stony corals kept in the aquarium.

Galaxea spp.

Galaxy, or crystal, coral

PROFILE

Available in green and brown colour variants, galaxy coral is an attractive species that should grace any reef aquarium able to fulfil its requirements. It has generally short tentacles, with tips of a different colour to that of the main structure.

WHAT size?

Massive colonies can form on natural reefs and their growth is similarly boundless in a home aquarium. It is difficult to take cuttings from this species, so allow it plenty of space to grow into, especially as its heavily armed sweeper tentacles can stretch over considerable distances.

WHAT does it eat?

You can offer this coral microplankton, but the majority of its nutritional requirements will be supplied by photosynthetic algae.

WHERE is it from?

Tropical Indo-Pacific.

WHAT does it cost?

★☆☆☆☆   ★★★☆☆

Prices have reduced over the years. Brilliant green forms are the most expensive, but even cultured ones are fairly reasonable. Lower price range is for brown individuals. Higher price is for larger or more attractive specimens.

▲ *Galaxy coral contributes movement and beauty to the aquarium, but watch out for those long and potentially lethal sweeper tentacles!*

HOW compatible with other invertebrates?

Allow plenty of space, both for the animal itself and its sweeper tentacles, otherwise it is capable of stinging and killing its neighbours.

WHAT water flow rate?

Moderate to strong currents. Lower flow rates may help to discourage the formation of sweeper tentacles.

HOW much light?

Moderate to strong illumination provided by T5 or metal-halide

SIMILAR SPECIES

Galaxy coral is unique in appearance and not easy to confuse with other species.

lighting is essential to meet the requirements of this coral.

WILL fish pose a threat?

Few reef-compatible species will bother the galaxy coral.

WILL it threaten fish?

This coral might harm fish if they contact its tentacles, but such events are rare.

WHAT to watch out for?

Try to choose specimens with the tentacles extended and avoid those with corallites devoid of tissue.

WILL it reproduce in an aquarium?

Yes, but rarely. There seems to be a minimum size at which certain species mature, but many aquarium specimens will not achieve this. Many corals and other invertebrates begin spawning as a result of external stimuli, such as the lunar cycle, but in effect these are absent from most home aquariums.

Goniopora stokesii

Flowerpot, or pompom, coral

PROFILE

To identify the genus correctly, look for small, almost spherical colonies, with long tubular polyps ending in a crown of 24 tentacles. Colours range from brown to metallic green. This is an unusual hard coral in that it is found in lagoon areas with low flow and turbid water. Place specimens on sand with plenty of room to expand. Although advances have been made in the husbandry of this coral, it remains difficult to maintain in the long term.

WHAT size?
Generally offered for sale at 5-10cm diameter. The skeleton of large individuals may achieve 40cm with a much wider polyp expansion, but most are smaller.

WHAT does it eat?
Contains symbiotic algae that provide a percentage of its nutritional requirements, but it demands more than this in the long term. Regular feedings of phytoplankton appear to result in a better success rate.

WHERE is it from?
Tropical Indo-Pacific.

WHAT does it cost?
★★☆☆☆
Most specimens are inexpensive but some, such as the metallic green individuals, are costlier.

▲ Trace elements and/or feeding may hold the key to long-term success.

HOW compatible with other invertebrates?
Can damage neighbouring corals, but if it comes into conflict with a more aggressive species, even the aquarist's speedy intervention may not prevent the encounter being fatal to the flowerpot coral. Once damaged, *Goniopora* is highly susceptible to protozoan and bacterial infections, which can strip the colony in hours.

WHAT water flow rate?
Low to moderate. Tolerates strong currents, but these may prevent full polyp expansion.

HOW much light?
T5 lighting of sufficiently high intensity or metal-halide lighting are essential, but do not guarantee success.

WILL fish pose a threat?
Few reef-safe fish attack the coral directly because of its sting. Anemonefish can cause damage or prevent polyp expansion when they 'move in' in the absence of a suitable host anemone.

WILL it threaten fish?
The well-developed stinging cells could harm fish if they contact the tentacles. However, most fish seem to have an innate wariness of stinging coral tentacles.

WHAT to watch out for?
Avoid specimens with partially or fully contracted tentacles. Ignore specimens with patches of exposed skeleton at all costs.

WILL it reproduce in an aquarium?
Healthy or newly imported specimens often develop miniature facsimiles that drop off onto the substrate. Spawning reports are very rare, probably due to the low rate of success in maintaining this coral long term.

SIMILAR SPECIES
There are several species of *Goniopora*, including the shorter-tentacled *Goniopora lobata,* which requires the same care as *G. stokesii*. It has pink and red forms and although not easy, is generally considered to be the hardier coral. *Alveopora* spp. are similar, but have fewer tentacles with a distinctive flattened tip.

Heliofungia spp.

Long-tentacle plate coral

PROFILE

A roughly circular skeleton, with a number of thin, vertically oriented plates, is home to an upper surface covered with long, round-ended polyps and a central mouth. Casual observers often mistake this species of stony coral for an anemone.

WHAT size?
Large specimens may reach 30cm across the skeleton, but the expanded tentacles make it appear much larger. Most specimens offered for sale are smaller.

WHAT does it eat?
Chunks of food can be placed directly on top of this animal, but it will also trap microplankton added directly to the water. It contains zooxanthellae.

WHERE is it from?
Tropical Indo-Pacific.

WHAT does it cost?
★★☆☆☆ ★★★☆☆
Large colourful specimens command the highest prices. Pink and fluorescent green variants are expensive.

▶ *Although often mistaken for an anemone, the plate coral generally does not enjoy being used as such by anemonefish. Even so, wild specimens often play host to a variety of commensal organisms.*

HOW compatible with other invertebrates?
Can sting other corals, but often fares worst in any competitive skirmishes, which can result in tissue being stripped from the skeleton. Give it plenty of room.

WHAT water flow rate?
Moderate to strong currents.

HOW much light?
Moderate to strong illumination from T5 or metal-halide lighting is essential to meet the requirements of this coral.

WILL fish pose a threat?
Certain fish seem to enjoy nipping at the tentacles or flesh of this coral. Some species of dwarf angelfish often find it irresistible. Often irritated by anemonefish that take it over as a surrogate anemone.

WILL it threaten fish?
The stinging tentacles have the potential to sting unwary fish, but most species seem to have an innate wariness of them.

WHAT to watch out for?
Avoid specimens with any exposed skeleton or deflated tentacles. Allow them to settle for a while before buying them.

WILL it reproduce in an aquarium?
Yes. It is reported to give off asexual planulae that can attach to rockwork. There, the disc grows on a small attachment point until it reaches a size where it can no longer support the weight of the colony, which then drops to the substrate.

SIMILAR SPECIES

This species can be confused with sand anemones (*Macrodactyla doreensis* and *Heteractis aurora*), due to its long tentacles. The skeleton could be confused with *Fungia* spp. plate corals, but they have much shorter and less dense tentacles.

Hydnaphora spp.

Elkhorn corals

PROFILE

Although it closely resembles branching *Acropora* and other small-polyp stony corals, this species is not particularly closely related to them. It is mostly available as cultured specimens, with brilliant fluorescent green coloration.

WHAT size?

Colonies of several metres in diameter are recorded on coral reefs, and given sufficient space to grow into, this coral has the potential to dominate almost any aquarium.

WHAT does it eat?

Contains zooxanthellae and will thrive in nutrient-poor systems intended for a predominance of small-polyp stony corals. Appreciates occasional feedings with microplankton.

WHERE is it from?

Tropical Indo-Pacific.

WHAT does it cost?

★☆☆☆☆ ★★☆☆☆
Prices have reduced with the prevalence of cultured individuals in the hobby. Price depends on the size and colour of the colony.

HOW compatible with other invertebrates?

This coral is highly aggressive and given its prodigious growth rate, presents a real threat to its neighbours. It will regularly sting them if its growth is not controlled.

WHAT water flow rate?

Easily tolerates moderate to strong currents.

HOW much light?

Enjoys strong lighting, so plenty of T5 lamps or metal-halide illumination is essential.

WILL fish pose a threat?

Few fish are tempted to tackle this coral.

WILL it threaten fish?

This coral should not harm fish, despite its potent sting.

SIMILAR SPECIES

SIMILAR SPECIES

There are several species of *Hydnaphora*. Some are almost exclusively encrusting, whereas others typically form branching colonies. All could be mistaken for similar growth forms of small-polyp stony corals, such as *Porites*, *Montipora* or *Acropora*.

WHAT to watch out for?

Specimens should show complete polyp coverage, with no white skeleton visible. Cultured specimens offer the best chances of success, but require strong illumination.

WILL it reproduce in an aquarium?

Can reproduce sexually in the home aquarium but this is rare. It usually spreads prolifically, encrusting first, then forming branches when the substrate has been conquered.

◀ *Green fluorescent colonies of* Hydnaphora *are beautiful, but often prove too aggressive to be maintained with small-polyp stony corals.*

Lobophyllia hemprichii
Spiny brain coral

PROFILE

A large-polyp stony coral that secretes long, sharp projections into its calcium carbonate skeleton as it grows, hence the common name. It varies in colour from brown to green to red. Some individuals have two distinct colours. The tentacles are normally retracted, but if food is placed on or near them they will slowly emerge to feed.

WHAT size?
Growth is slow but steady in the aquarium and colonies measuring over 30cm across are not unattainable.

WHAT does it eat?
Primarily photosynthetic, but will feed on a wide variety of meaty foods, such as chopped shellfish and mysis and even flaked foods. If food is placed directly onto the surface, the tentacles emerge after 5-10 minutes and then more food can be given directly onto these sticky projections.

WHERE is it from?
Tropical Indo-Pacific.

WHAT does it cost?
★☆☆☆☆   ★★☆☆☆
One of the least expensive corals commonly imported from the tropical Indo-Pacific. Price will depend on colour and size.

Aquarists like specimens with red and green fluorescent pigments.

HOW compatible with other invertebrates?
Although capable of using its tentacles to sting its neighbours, this species often proves reluctant to defend itself against more aggressive neighbours and needs ample space to expand and grow into.

WHAT water flow rate?
Low to moderate. This coral does not enjoy strong flow, which can present an obstacle to proper tissue expansion.

HOW much light?
Will thrive and grow under T5 and metal-halide lighting. Can be maintained under T8 lamps, given sufficient wattage.

SIMILAR SPECIES
This coral closely resembles the open brain coral *(Trachyphyllia geoffryi)*. At present, there are several restrictions concerning the trade of *T. geoffryi*.

WILL fish pose a threat?
Dwarf angelfish are notorious nippers of large-polyp stony corals, but some species are worse than others. The coral beauty *(Centropyge bispinosus)* is the safest. Otherwise relatively trustworthy true angelfish, such as the majestic *(Pomacanthus navarchus)* and regal *(Pygoplites diacanthus),* may be unable to resist having a nip at this coral.

WILL it threaten fish?
Is capable of stinging fish, particularly slow-moving species, but this seldom proves fatal or causes the victim long-lasting problems.

WHAT to watch out for?
Avoid specimens with visible white skeleton. The base of the coral is often home to several species of encrusting organism. Make sure these are alive, as certain animals, such as sponges, sometimes die off and pollute the water.

WILL it reproduce in an aquarium?
This coral can and will spawn where conditions allow.

Montipora capricornis

Plating montipora

PROFILE

A beautiful species of small-polyp stony coral that grows in thin, laminar plates, often forming cup-shaped colonies. Red forms are highly prized, but commonly offered for sale as increasing numbers of fragged colonies become available.

WHAT size?
Almost no upper limits to its size in the aquarium. It grows quickly and, left unchecked, can occupy large spaces.

WHAT does it eat?
Traps microplankton from the water column, but relies on its photosynthetic algae to provide the majority of its nutritional requirements in the aquarium.

WHERE is it from?
Tropical Indo-Pacific.

WHAT does it cost?
★☆☆☆☆   ★★★★☆
Price depends on size, colour and point of origin. There is little point in buying wild-collected specimens, as cultured or fragged colonies are widely available.

▶ *The leaflike shapes of these plating* Montipora *offer something different for the reef aquarist. However, the coral can be brittle; take care when performing in-tank maintenance close to it.*

HOW compatible with other invertebrates?
This coral is not an especially aggressive species and rather than posing any significant threats in its own right, will usually suffer from the aggressive tendencies of its neighbours.

WHAT water flow rate?
Moderate to strong indirect flow.

HOW much light?
Moderate to high. Many aquarists cultivate this species using T5 or multiple T8 fluorescents, but metal-halide lighting often gives the best results.

WILL fish pose a threat?
Few, if any, fish generally considered to be reef-safe are likely to bother this coral.

WILL it threaten fish?
No. This is an entirely benign species.

SIMILAR SPECIES
One of a number of small-polyp stony corals that can form plating colonies. Many of these are not available due to CITES restrictions, but *Echinopora* is similar and available with some regularity. This coral has distinct rounded lumps (corallites) on its surface, absent in *M. capricornis*, and in most cases, a drabber appearance.

WHAT to watch out for?
Avoid specimens showing any patches of bare white skeleton. Polyps are unlikely to be visible in this species.

WILL it reproduce in an aquarium?
Sexual reproduction is possible but very rare.

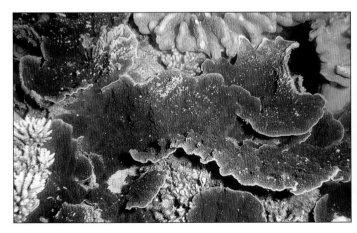

Platygyra spp.

Maze coral

PROFILE

Named for the meandering shape of its polyps, this is a beautiful coral that is often available with metallic pigments. Specimens can be of a single colour, but more usually have brown outer margins with green centres.

WHAT size?
Specimens are generally offered for sale at 10-15cm across, but slowly grow much larger.

WHAT does it eat?
Contains symbiotic algae, but will also accept fine particulate foods, such as microplankton substitutes. Feeds in the same way as *Favia* spp.

WHERE is it from?
Tropical Indo-Pacific.

WHAT does it cost?
★☆☆☆☆ ★★★☆☆
Price is determined by size, colour and point of origin. Cultured specimens are available but are usually small, albeit with strong colours.

SIMILAR SPECIES
Not many corals are confused with this species, at least not amongst regularly imported species. *Diploria* and *Colpophyllia* spp. both have a similar meandering polyp form.

▲ *The contrast of brown with fluorescent green pigments make this a stunning and popular coral.*

HOW compatible with other invertebrates?
This is not an overly aggressive species of stony coral, although the tentacles are well equipped to defend the animal where necessary. It is vulnerable to species with long tentacles bearing a potent sting.

WHAT water flow rate?
Prefers moderate currents.

HOW much light?
Must be provided with strong illumination. Brown specimens are generally tolerant of lower light intensity, but metallic-green ones require higher levels.

WILL fish pose a threat?
Dwarf angelfish regularly threaten this coral by nipping at its tissue. However, this does depend on the fish species involved; the coral beauty

(*Centropyge bispinosa*) is fairly trustworthy, whereas the lemonpeel (*C. flavissimus*) or flame dwarf angel (*C. loricula*) are commonly problematic.

WILL it threaten fish?
No. Has the potential to sting fish when the tentacles are everted, but this rarely happens.

WHAT to watch out for?
Avoid specimens with any signs of white skeleton or receding tissue.

WILL it reproduce in an aquarium?
Spawning has been recorded for aquarium specimens but is not common.

Plerogyra sinuosa

Bubble coral

The bubble coral is a beautiful species, readily identified by its large, distinctive polyps. These are usually pink to white in colour. The rare green form is highly sought after. Despite its somewhat delicate appearance, it is relatively hardy and easy to maintain.

WHAT size?
Colonies of several metres in diameter are recorded on coral reefs, and given sufficient space to grow into, this coral has the potential to dominate almost any aquarium.

WHAT does it eat?
Contains zooxanthellae, but has a liking for regular feeding with chunks of shellfish, brineshrimp and mysis. Placing these items directly onto the animal will stimulate it to extend pointed stinging tentacles from between the large bladderlike polyps.

WHERE is it from?
Tropical Indo-Pacific.

WHAT does it cost?
★★☆☆☆   ★★★★☆
The price for a colony has risen over recent years and it is now one of the more expensive large-polyp stony corals available with any degree of regularity. Price depends on size and colour.

HOW compatible with other invertebrates?
Try to give this coral plenty of room. Not only is it aggressive in its own right, but it will also be stung by its neighbours, resulting in tissue die-off.

WHAT water flow rate?
Low to moderate. Tolerates strong currents, but these often prevent full polyp expansion.

HOW much light?
Moderate to strong lighting. Thrives best under metal-halide or T5 lights, but there are reports of good success with sufficient numbers of T8 lamps. Do not locate the coral directly beneath metal-halides rated higher than 150W.

WILL fish pose a threat?
Some otherwise reef-safe fish may occasionally pick at the large polyps, but most avoid them. However, the presence of food on or around the coral's tentacles sometimes invites too much attention from fish, with the result that the polyps remain retracted.

SIMILAR SPECIES
Two other regularly imported species, both members of the genus *Physogyra*, are also known as bubble corals. *P. lichtensteini* is usually fairly flat in shape, with small pearl-like bubbles, hence its common name of pearl, or grape, bubble coral. *P. flexuosa* is commonly called the octobubble and has small irregular-shaped bubbles arising from the polyps.

WILL it threaten fish?
The tentacles' potent sting is potentially hazardous, but under normal circumstances, few fish stray close enough to be harmed.

WHAT to watch out for?
As with all corals, make sure that the polyps are expanded before buying them. Signs of the skeleton showing through may suggest an ailing specimen.

WILL it reproduce in an aquarium?
Sexual reproduction has been reported, but is not common, perhaps because colonies need to reach a particular size before becoming sexually mature.

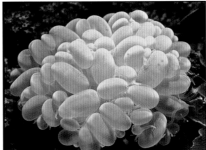

◀ *Some aquarists love the characteristic polyp form of the bubble coral, whereas others find it unappealing.*

Porites cylindrica

Branching porites

Available in many colour varieties, this small-polyp stony coral forms beautiful branching colonies with a robust appearance. Yellow cultured specimens are stunning and commonly offered for sale.

WHAT size?

Will easily grow to over 30cm in diameter in the home aquarium, but perhaps less quickly than other small-poly stony corals due to its more robust skeletal form.

WHAT does it eat?

Traps microplankton from the water column, but relies on zooxanthellae to provide the majority of its nourishment in the aquarium.

WHERE is it from?

Tropical Indo-Pacific.

WHAT does it cost?

★☆☆☆☆ ★★★★☆

Price depends on size, colour and point of origin. *Porites* colonies tend to grow more slowly than many other small-polyp stony coral species and can therefore command slightly higher prices.

◄ *Porites cylindrica can be distinguished from similar species, such as Montipora digitata, by close scrutiny of the corallites. However, the husbandry for both species is much the same.*

HOW compatible with other invertebrates?

The only aggression you are likely to see is when this coral grows close to another species of small-polyp stony coral. Then it is likely to do battle, using stinging cells to attempt to kill its neighbour, or at least discourage its growth.

WHAT water flow rate?

Moderate to strong indirect flow.

HOW much light?

Prefers strong illumination. It has been suggested that brightly coloured morphs, such as purple or blue, do well under higher-rated Kelvin lamps, but many aquarists report success with multiple T5 or metal-halide lamps with standard 13,000K-16,000K bulbs.

WILL fish pose a threat?

Few fish bother *Porites*.

WILL it threaten fish?

No. Where fish are concerned, this is an entirely benign species.

WHAT to watch out for?

Avoid specimens with bare white skeleton showing. Healthy colonies often appear to have a covering of fine, feltlike polyps.

WILL it reproduce in an aquarium?

Can reproduce through polyp bale out. Living polyps are dropped from the parent colony and are able to form new colonies where they settle.

SIMILAR SPECIES

Very similar to *Montipora digitata*. Apart from differences in the structure of the skeleton, *M. digitata* tends to grow more quickly and have more fragile branches in aquarium specimens, especially in lower-flow areas of the aquarium. The red-orange form is easy to maintain and commonly available as fragged specimens.

Seriatopora hystrix

Bird's nest coral

PROFILE

One of around five species of small-polyp stony corals found in this genus. Despite their delicate appearance, all will thrive under strong illumination and in good-quality, nutrient-poor water. Pink morphs are the most desirable, but all are beautiful and grow rapidly.

WHAT size?
Most colonies will be modest in size, but they can reach over 30cm across.

WHAT does it eat?
Most of its requirements will be satisfied by its photosynthetic algae, but it will also feed on fine particulate material, such as microplankton, from the water column.

WHERE is it from?
Indo-Pacific.

WHAT does it cost?
★☆☆☆☆   ★★★★☆
Frags are inexpensive and represent the best specimens for the hobbyist. Wild-collected colonies and individuals cultured in tropical climes are expensive. Both may experience colour change in the aquarium. The price range covers frag to larger, coloured colony.

▲ *To maintain the pink coloration provide a very low-nutrient system.*

HOW compatible with other invertebrates?
Not a particularly aggressive coral. Give it plenty of protective space to grow into.

WHAT water flow rate?
Strong, indirect currents.

HOW much light?
Multiple T5 lamps or metal-halide lighting are particularly helpful in maintaining strong colours. Some aquarists favour the use of higher Kelvin-rated lamps (16,000K or 20,000K) that may stimulate the retention of UV reflective pigments.

WILL fish pose a threat?
Few reef-compatible fish will bother the bird's nest coral.

WILL it threaten fish?
This coral should not harm fish.

WHAT to watch out for?
White-skeleton or excessively pale specimens are best ignored. Be aware that brightly coloured specimens collected from the wild have a lower chance of retaining their attractive appearance than similar specimens purchased as aquarium-grown frags.

WILL it reproduce in an aquarium?
Seriatopora can reproduce in the aquarium by dropping living polyps that settle and form new colonies. It is easily fragged.

SIMILAR SPECIES
Other species belonging to the genus *Seriatopora* have the distinctive arrangement of polyps in uniform lines. Otherwise, there is at least one species of *Acropora* with very pointed tips and tiny polyps.

Stylophora pistillata
Cat's paw coral

PROFILE

A beautiful coral, available in pink and purple forms. The rounded appearance to the branching structure of the calcareous skeleton is distinctive. This small-polyp stony coral demands stable water parameters of sufficiently high quality.

WHAT size?
30cm is a large colony, even on natural reefs. Most specimens offered for sale are much smaller than this.

WHAT does it eat?
Has symbiotic algae, but will also accept microplankton. Probably absorbs amino acids and similar molecules directly from the water column.

WHERE is it from?
Tropical Indo-Pacific.

WHAT does it cost?
★☆☆☆☆ ★★★★☆
Cuttings or frags are modestly priced, depending on their coloration and size. The same is true of wild-collected specimens or those cultured in tropical climes, but these latter examples are likely to be more expensive.

▶ Stylophora pistillata *is regularly available as a cultured specimen grown in the Indo-Pacific. Colours vary significantly, from golden brown to brilliant pink.*

HOW compatible with other invertebrates?
This is not a particularly aggressive coral and will often be the loser in any competitive disagreements with its neighbours.

WHAT water flow rate?
Will tolerate a wide range of currents, from almost none to quite vigorous. The amount of flow aimed towards the colony will influence its growth form.

HOW much light?
Only aquarists using multiple T5 lamps or metal-halide systems should consider keeping this coral. The most desirable specimens are the pink and purple individuals, which are typically found in shallow water

SIMILAR SPECIES
A less squat, more highly branched species of *Stylophora* is sometimes offered for sale. Many small-polyp stony corals also bear a resemblance to cat's paw coral, including *Montipora* spp., *Porites* spp. and *Acropora* spp.

and abundant light. To retain their colours in the aquarium it is important to replicate these conditions.

WILL fish pose a threat?
Rogue fish may occasionally nibble at this coral, but most species ignore it entirely.

WILL it threaten fish?
This coral should not harm fish.

WHAT to watch out for?
Ignore excessively pale specimens, as they may have lost their photosynthetic algae. Check for snow-white patches, present when the living tissue has receded from the skeletal material.

WILL it reproduce in an aquarium?
Potentially, but rarely reported.

Tubastrea spp.

Sun, or orange, coral

With its large polyps on a pink-orange background, the sun coral is a beautiful animal. Although once thought to shun light, it is easily trained to emerge during the day and is currently considered to be a fairly easy coral to maintain, given good water quality.

WHAT size?
Its expansion in the aquarium is usually limited only by the available space to grow into.

WHAT does it eat?
Feed regularly, perhaps daily, for the best results. Mysis shrimp is large enough to be ingested. Tempt the polyps to expand by placing food directly onto the coral or by adding zooplankton substitutes, such as frozen marine copepods or oyster eggs, directly to the water column. When everted, after 10-20 minutes, feed the sticky polyps individually from a syringe or pipette. Try to feed as many polyps as possible. Soon, the polyps will expand when they detect their usual food in the water, say, when fish are fed.

WHERE is it from?
Tropical Indo-Pacific.

WHAT does it cost?
★☆☆☆☆   ★★☆☆☆
Price depends on the size of the colony.

Sun corals need regular feeding if they are to survive and grow.

HOW compatible with other invertebrates?
Will be stung by aggressive sessile neighbours, but the greatest threat is from free-living species, such as shrimps and crabs, that raid the polyps for their food, even if it is half ingested. Placing a protective barrier over the colony, such as a plastic drinks bottle with the bottom removed, can help prevent this happening during feeding. Be aware that food introduced for this coral will create waste; consider the impact of this on other invertebrates in terms of water quality.

WHAT water flow rate?
Low to moderate. Tolerates strong currents, but these take the food away. Turning off pumps temporarily when feeding helps.

HOW much light?
Site the coral low down in the aquarium or in a shaded spot.

WILL fish pose a threat?
This coral will be irritated by fish stealing its food. Few fish directly attack the polyps, but some rogues may target them.

WILL it threaten fish?
The coral's stinging cells are seldom, if ever, harmful to fish.

WHAT to watch out for?
Avoid specimens with patches of white skeleton amid the orange or pink base colour. Choose healthy specimens from dealers who have begun the feeding process.

WILL it reproduce in an aquarium?
Yes, will produce planula larvae through asexual means. These settle and grow on hard surfaces, such as rockwork or glass.

SIMILAR SPECIES
There are similar species with large polyps and an absence of zooxanthellae. *Dendrophyllia* spp. have a yellow base colour. They are found on the same specimen as *Tubastrea* spp., making a beautiful display. *T. micrantha* is a black-dark green species, often forming branching colonies. Although slightly more temperamental than the orange/pink forms, it requires the same care.

Turbinaria crater

Crater, or table, coral

PROFILE

This coral is a good species for anyone wishing to start keeping stony corals. Its characteristic growth form is attractive, and healthy specimens are always in demand. Fortunately, it is easy to obtain.

WHAT size?
Large individuals may reach 30cm or more in diameter and a similar height. Specimens are available for sale at 5-10cm diameter and are easy to place in the aquarium.

WHAT does it eat?
Zooxanthellae supply the majority of the coral's nutritional requirements in the absence of supplemental feeding. The polyps will accept small particulate foods, such as frozen marine copepods, etc. Using a syringe, squirt the food onto the polyps or drop it into an area of current and allow it to disperse.

WHERE is it from?
Tropical Indo-Pacific.

WHAT does it cost?
★☆☆☆☆ ★★☆☆☆
For a stony coral this is an inexpensive species.

▶ *Beginners to keeping stony corals will find specimens of* Turbinaria *with relatively large polyps the best prospect to care for.*

HOW compatible with other invertebrates?
Not particularly aggressive. May not fare well against competition from its neighbours. Give it plenty of space for growth and to protect it.

WHAT water flow rate?
Moderate currents are usually sufficient to keep the 'crater' free of organic material.

HOW much light?
Will survive under T8 lights of sufficient number and wattage, but T5 or metal-halide illumination provides a better chance of long-term survival.

WILL fish pose a threat?
Few fish bother this coral. Dwarf angelfish may occasionally take a peck at the polyps.

WILL it threaten fish?
No.

WHAT to watch out for?
Due to their shape, certain specimens may retain detritus in the crater, especially where

SIMILAR SPECIES
At least two more species of *Turbinaria* are commonly offered for sale. *T. pagoda* is very similar in polyp form to *T. crater*, but grows in tall columns or vertical plates. *T. mesenterina* is a yellow form known as 'scroll coral' that thrives under strong illumination. It has small polyps that may or may not expand in the aquarium. Either condition seems to be acceptable for its long-term survival.

▲ *Golden, or scroll,* Turbinaria *is slightly more challenging to keep than the larger-polyp species.*

there is insufficient flow. If not removed regularly, this can lead to tissue die-off. Healthy specimens should have tissue all over their skeleton, including the base, which is often shaded by the cup part of the coral.

WILL it reproduce in an aquarium?
Spawning events have been recorded but are rare.

Swaying branches

Known as sea whips, sea plumes and sea fans, the gorgonians are often highly branched animals with beautiful polyps. The flesh and polyps extend over a thin, wirelike 'skeleton', and aquarists often look for signs of exposed skeleton as an indication of a damaged or ailing specimen. For convenience we can divide gorgonians into two groups: those that contain photosynthetic algae and those that do not. In general, the photosynthetic species are much easier to maintain in the reef aquarium as they tend to enjoy similar conditions to those required by other photosynthetic organisms. However, recent improvements in the types and availability of foods intended for filter-feeding marine organisms has meant that species hitherto considered difficult to care for in the long term can now thrive if well cared for by the aquarist.

Price guide

★	£12 – 15
★★	£15 – 25
★★★	£25 – 35
★★★★	£35 – 40
★★★★★	£40 – 50

PROFILE

With its beautiful combination of blue polyps and orange horny skeleton, this sea fan attracts many aquarists. However, to have any chance of success, it must be regularly fed with the correct diet. Given recent advances in the development of invertebrate foods, it is easier to keep, but still represents a significant challenge to hobbyists.

WHAT size?
Large specimens reach 50cm across the fan, but most aquarium specimens achieve more modest proportions, often never growing any larger than when first acquired.

WHAT does it eat?
Offer a variety of diets in a number of size ranges. Phytoplankton, microplankton and larger zooplankton substitutes are all very useful.

WHERE is it from?
Tropical Indo-Pacific.

WHAT does it cost?
★★☆☆☆   ★★★★★
Relatively expensive.

Blue polyp gorgonians in a dealer's aquarium may not necessarily show the polyp opening or expansion seen in this individual. Check for healthy-looking branches and avoid specimens with evidence of tissue recession.

Acalycigorgia spp. (*Acanthogorgia* spp.)

Blue polyp sea fan

HOW compatible with other invertebrates?

Stinging corals and anemones may harm the sea fan. Of the free-living animals, shrimps and crabs may target the polyps if they contain food. Otherwise, few reef-safe invertebrates will bother this animal.

WHAT water flow rate?

Strong flow is required to stimulate the polyps to expand.

HOW much light?

Independent of light, but may be reluctant at first to expand its polyps when under direct illumination from bright lights.

WILL fish pose a threat?

The main potential threat is from fish nipping at polyps as they close around food items. Few fish pose any direct threat.

WILL it threaten fish?

No.

WHAT to watch out for?

Be sure to observe this coral with the polyps expanded before buying it. Check for any signs of the wirelike exposed skeletal material, which can indicate an ailing specimen.

SIMILAR SPECIES

Several non-photosynthetic sea fans are offered for sale with some degree of regularity. All require feeding, but in most cases the chances of maintaining them successfully long term are limited.

WILL it reproduce in an aquarium?

No, at least not until more aquarists report success in keeping this animal alive long term.

Diodogorgia nodulifera

Colourful sea rod

Aquarists are often drawn to this inexpensive, beautiful and variable gorgonian, with its beautiful snow-white polyps. Historically, it has been difficult to maintain for any length of time in captivity, but with the wealth of specialist invertebrate diets currently at the aquarist's disposal, there are more reports of success with this species.

WHAT size?
Usually acquired at 10-15cm in height. Grows to about 30cm in its natural environment.

WHAT does it eat?
Daily feedings with particulate diets, including zooplankton substitutes and phytoplankton, have been successful in the maintenance of this species.

WHERE is it from?
The Caribbean.

WHAT does it cost?
★☆☆☆☆
Inexpensive.

▶ *Aquarists should ideally seek specimens that are in good general health, with the polyps expanded.*

HOW compatible with other invertebrates?
Presents no problems to other invertebrates.

WHAT water flow rate?
Enjoys moderate to strong indirect currents that help to keep it free from detritus and encourage polyp expansion.

HOW much light?
Contains no symbiotic algae. It will open under direct lighting, but in aquariums with excess levels of nitrate or phosphate, it may suffer from algae growing on its surface. This causes the polyps to remain retracted. Placing the sea rod in lower light or even complete shade can alleviate the problem, at least until the underlying water quality issue is addressed.

WILL fish pose a threat?
Fish, including dwarf angelfish, may nibble at polyps, particularly if they hold particulate food at the time. Certain fish may also target algae that fouls the gorgonian, but this tends only to occur in poorly maintained specimens.

WILL it threaten fish?
No.

WHAT to watch out for?
The two most common colour varieties are yellow and red, both with white or transparent polyps. Avoid specimens showing any powdery deposits beneath them, as this is a sign of ill-health. Similarly, ignore pieces with any tissue recession. Be aware that the amount of food required to sustain such animals will often put large demands on aquarium filtration and may compromise water quality.

WILL it reproduce in an aquarium?
Unlikely.

SIMILAR SPECIES
Several species of gorgonian are similar to *Diodogorgia,* including the emperor gorgonian *Echinogorgia* sp. (page 85). The aquarium requirements for most of the larger-polyped, non-photosynthetic gorgonians are similar.

Echinogorgia sp.

Emperor gorgonian

PROFILE

This maroon gorgonian, with beautiful deep-golden polyps, has become a fairly regular import over recent years. However, it does not photosynthesise and it is best left to experienced aquarists to care for. Even so, there are many aquarists reporting long-term success with this animal.

WHAT size?
Can grow to over 60cm in diameter, but most aquarium specimens never achieve this. Individuals offered for sale are usually less than 45cm.

WHAT does it eat?
Being devoid of zooxanthellae, this coral requires regular daily feedings with plankton substitutes, such as frozen marine copepods, oyster eggs and similar. The greater the variety the better, and in good quantities.

WHERE is it from?
Tropical Indo-Pacific.

WHAT does it cost?
★★☆☆☆ ★★★★★
Price will depend on size.

▶ *The increased availability and diversity of zooplankton substitutes available to aquarists have meant that reef aquarists can maintain beautiful non-photosynthetic species such as this long term.*

HOW compatible with other invertebrates?
Can be irritated by close proximity to disc anemones (*Discosoma* spp., *Rhodactis* spp. and *Ricordea* spp.) or any corals with stinging tentacles. Often imported with commensal brittle starfish in residence, sometimes in large numbers. These are harmless.

WHAT water flow rate?
Strong flow is essential for this gorgonian to be able to shed its mucus film. If provided by large-aperture circulation pumps, the flow should be direct; when standard powerheads are used, provide indirect flow.

SIMILAR SPECIES
Several species of branching, non-photosynthetic gorgonians are offered for sale, but most are as difficult to care for as the emperor.

HOW much light?
Prefers lower light levels, although the polyps will often expand beneath metal-halide illumination.

WILL fish pose a threat?
This gorgonian should not be bothered by any reef-safe aquarium fish species.

WILL it threaten fish?
No.

WHAT to watch out for?
Avoid specimens with powdery deposits around the base, which means they are shedding body tissue and indicates an unhealthy animal. Always choose specimens with full polyp expansion.

WILL it reproduce in an aquarium?
No.

Eunicea spp.

Warty sea rod

PROFILE

Large branches of robust appearance, sometimes with very large (up to 10mm-diameter) polyps, characterise the sea rods. The warty species from the genus *Eunicea* have a knobbly appearance. They contain photosynthetic pigments and are therefore amongst the easier gorgonians to maintain in the marine aquarium.

WHAT size?
Specimens for sale can be up to 45cm tall. They can grow to 60cm or more.

WHAT does it eat?
Will pluck small particles from the water column, but the polyps contain zooxanthellae and these provide the majority of the animal's needs.

WHERE is it from?
The Caribbean.

WHAT does it cost?
★★☆☆☆
★★★★★
Large specimens can be expensive depending on the distance that they have had to travel.

HOW compatible with other invertebrates?
Can be irritated by close proximity to disc anemones (*Discosoma* spp., *Rhodactis* spp. and *Ricordea* spp.) or any corals with stinging tentacles.

WHAT water flow rate?
Provide moderate flow. Indirect currents are best.

HOW much light?
Try to offer this gorgonian moderate to strong illumination.

SIMILAR SPECIES
The genus *Eunicea* contains a few species that can make their way into the aquarium hobby. The husbandry is similar for each. They can be confused with the encrusting star polyp *Briareum asbestinium* that often overgrows dead gorgonians. This is sometimes sold as purple thumb gorgonian.

Metal-halide is best, but it will tolerate multiple T8 or T5 lamps.

WILL fish pose a threat?
Very few reef-safe fish will bother this animal. Dwarf angelfish may occasionally nibble at the soft flesh, but this will depend on the individual fish concerned.

WILL it threaten fish?
No.

WHAT to watch out for?
Avoid specimens with any sign of the exposed internal skeleton, as they can deteriorate quickly.

WILL it reproduce in an aquarium?
Spawning is not out of the question. Cuttings can be taken with a good chance of success.

◀ *The lumpy texture created by the retracted polyps of this gorgonian are characteristic of the genus.*

Muricea sp.

Silver tree gorgonian

PROFILE

A commonly available photosynthetic gorgonian, tentatively identified as a species of *Muricea*. It has long branches that taper slightly along their length to a point. The branches are silvery grey and the polyps brown. It has a rough texture.

WHAT size?
Specimens offered for sale can be up to 25cm tall. They can grow to 60cm or more.

WHAT does it eat?
Accepts small particles from the water column, but the polyps contain zooxanthellae and these will provide the majority of the gorgonian's needs in the aquarium.

WHERE is it from?
Tropical Atlantic – Caribbean.

WHAT does it cost?
★☆☆☆☆ ★★☆☆☆
Inexpensive.

SIMILAR SPECIES
Many species of branching gorgonian can be confused with the silver tree, but the combination of grey-silver base colour, brown polyps and a rough texture distinguish this species from most others.

▲ This gorgonian has small and delicate polyps that harbour the photosynthetic symbionts.

HOW compatible with other invertebrates?
Can be irritated by close proximity to disc anemones (*Discosoma* spp., *Rhodactis* spp. and *Ricordea* spp.) or any corals with stinging tentacles.

WHAT water flow rate?
Provide moderate flow. Indirect currents are best.

HOW much light?
Try to offer this gorgonian moderate to strong illumination. Metal-halide is best, but it will tolerate a number of T5 lamps of sufficient wattage.

WILL fish pose a threat?
Few fish show any interest in this animal.

WILL it threaten fish?
No.

WHAT to watch out for?
Avoid specimens with any sign of exposed, wirelike, internal skeleton. Such specimens can be salvaged, given good water quality and a suitable place in a well-run aquarium, but do not undertake this challenge when it is possible to buy healthy specimens.

WILL it reproduce in an aquarium?
Spawning has not been recorded. Cuttings can be taken with confidence of success.

▲ The silver tree gorgonian has a highly flexible structure and is tolerant of high flow. However, it is rarely shipped with a substantial rock base and can be difficult to site unless permanently fixed to rockwork.

Pseudoplexaura spp.

Porous sea rod

PROFILE

Although smooth in appearance, the thickset branches of the porous sea rods are home to large polyps that retract into characteristic oval-shaped recesses in the tissue. The animal's high centre of gravity and relatively small anchor point can make it difficult to site. It is best fixed using a specialist putty or glue.

WHAT size?
Specimens offered for sale are generally less than 45cm tall, but this gorgonian can reach 2m tall in the tropical Atlantic reef zones it calls home.

WHAT does it eat?
Can trap small particles, including microplankton and brineshrimp larvae, from the water column, but its photosynthetic algae tend to provide the greater part of its nutrition in the aquarium.

WHERE is it from?
Tropical Atlantic – Caribbean.

WHAT does it cost?
★★☆☆☆ ★★★★★
Price is determined solely by the size of the colony.

▶ *Specimens showing full polyp expansion are beautiful to behold, and this is also an indication that the specimen is in good health.*

HOW compatible with other invertebrates?
Can be irritated by close proximity to disc anemones (*Discosoma* spp., *Rhodactis* spp. and *Ricordea* spp.) or any corals with stinging tentacles.

WHAT water flow rate?
Moderate to strong flow is an important ingredient of success with these animals.

HOW much light?
Strong illumination is essential for the long-term maintenance of this gorgonian. Provide T5 or metal-halide lighting.

WILL fish pose a threat?
Few fish tend to bother this animal, but dwarf angelfish, true angelfish and similar browsing species may nibble at it.

WILL it threaten fish?
No.

WHAT to watch out for?
Avoid specimens showing signs of deterioration in the base tissue, as indicated by exposed skeleton. Powdery deposits around the base also show that a specimen may be ailing.

WILL it reproduce in an aquarium?
Spawning possible. This coral can be fragged with care.

Pseudopterogorgia bipinnata

Purple frilly gorgonian

PROFILE

A stunning gorgonian that ranks as one of the easiest of this group to maintain in the aquarium. It is also known as the purple sea plume.

WHAT size?
Given space in which to grow, the animal can reach over 45cm tall, with each branching arm measuring 10-12cm.

WHAT does it eat?
Relatively easy to maintain, because the polyps contain photosynthetic algae. It should also filter small amounts of particulate material from the water column if offered.

WHERE is it from?
Tropical Indo-Pacific.

WHAT does it cost?
★★☆☆☆ ★★★☆☆
Price depends on size.

SIMILAR SPECIES
Could be confused with other species from the genus *Pseudopterogorgia*, but unlikely to resemble closely any species of gorgonian commonly imported for the aquarium trade.

▲ *With the polyps retracted, the vivid purple pigments can be clearly seen. Choose specimens with no tissue recession or areas of discoloration.*

HOW compatible with other invertebrates?
Can be irritated by close proximity to disc anemones (*Discosoma* spp., *Rhodactis* spp. and *Ricordea* spp.) or any corals with stinging tentacles.

WHAT water flow rate?
Prefers strong flow, preferably indirect, unless supplied by specialised flow pumps that deliver high-volume, low-pressure currents.

HOW much light?
Requires strong illumination, so best maintained under T5 or metal-halide lighting.

WILL fish pose a threat?
Very few reef-safe fish will bother this animal.

WILL it threaten fish?
No.

WHAT to watch out for?
Avoid specimens in which the purple tissue has receded from the thin wirelike skeletal tissue.

WILL it reproduce in an aquarium?
Can be fragged. Spawning events have been commonly reported but, given its planktonic larval stages, the chances of any young surviving are impossibly small.

▲ *When the polyps are expanded, the gorgonian takes on an entirely different appearance, often appearing completely brown. Specimens with complete polyp expansion are usually in excellent health and being provided with the right conditions.*

Rumphella spp.

Bushy gorgonian

Long, beige-brown branches, unable to support their own weight upright, are characteristic of this genus of photosynthetic gorgonian. It is not commonly imported and those that do make it into dealers' tanks are sometimes viewed as difficult to maintain. However, being photosynthetic it is easier to keep in the average reef aquarium than non-photosynthetic species, which can be difficult to feed.

WHAT size?
Can grow to over one metre across the branches. Aquarium specimens are offered at 15-45cm.

WHAT does it eat?
Will accept small particles of food from the water column, but its polyps contain photosynthetic algae that provide the majority of its food requirements in a home aquarium.

WHERE is it from?
Tropical Indo-Pacific.

WHAT does it cost?
★★☆☆☆   ★★★★☆
Can be pricey depending on size. The price range shown here reflects small specimens to larger colonies.

▲ *Examples of this group of photosynthetic gorgonians can prove hardy and long-lived in the reef aquarium and are undeniably attractive.*

HOW compatible with other invertebrates?
Can be irritated by close proximity to disc anemones (*Discosoma* spp., *Rhodactis* spp. and *Ricordea* spp.) or any corals with stinging tentacles.

WHAT water flow rate?
Strong flow is essential to enable the animal to shed its mucus film. Flow should be direct if supplied by large-aperture circulation pumps, but indirect when standard powerheads are used.

HOW much light?
Moderate to strong illumination is very important. T5 of sufficient quantity/wattages are a reasonable alternative to the preferred metal-halide lamps.

WILL fish pose a threat?
Very few reef-safe fish will bother this animal.

WILL it threaten fish?
No.

WHAT to watch out for?
Specimens do not like to experience algal growth on or near them. The former is sometimes a problem in systems with higher-than-acceptable levels of nitrate and/or phosphate. The algae tends to grow on the mucus secreted by the animal, but can be shed given sufficient water flow. Avoid specimens with exposed skeleton.

WILL it reproduce in an aquarium?
Spawning is not out of the question and cuttings can be taken with a good chance of success.

SIMILAR SPECIES
There are several species/growth forms of *Rumphella* for which the husbandry is identical. Several similar gorgonians are available to hobbyists, albeit sporadically. Check out the expanded polyps. If they are any colour other than brown or golden, then the chances are that the animal does not contain zooxanthellae.

Swiftia exserta

Tangerine gorgonian

PROFILE

A highly branched gorgonian with a delicate appearance. This species requires specialist care and a certain amount of dedicated assistance on the part of the aquarist if it is to survive. It is one of the more commonly imported, non-photosynthetic gorgonians.

WHAT size?
Can be 30cm tall at sale and reaches up to 45cm.

WHAT does it eat?
Needs regular feeding with plankton substitutes. Brineshrimp adults and larvae, frozen marine copepods and similar products are all suitable and essential for this animal's long-term maintenance.

WHERE is it from?
Tropical Atlantic – Caribbean.

WHAT does it cost?
★☆☆☆☆
Inexpensive. Price depends on travelling distance to market.

SIMILAR SPECIES
The genus *Swiftia* contains several species, but few are likely to be confused with the tangerine gorgonian.

HOW compatible with other invertebrates?
Can be irritated by close proximity to disc anemones (*Discosoma* spp., *Rhodactis* spp. and *Ricordea* spp.) or any corals with stinging tentacles. Ornamental shrimp may attempt to remove food from the polyps, sometimes damaging these in the process.

WHAT water flow rate?
Provide moderate flow. Indirect currents are best.

HOW much light?
Light-independent. Sometimes initially reluctant to expand its polyps during the aquarium's daylight period.

WILL fish pose a threat?
Few fish bother this gorgonian, but it can be threatened by fish attempting to remove food from its polyps.

WILL it threaten fish?
No.

WHAT to watch out for?
Avoid specimens with tissue recession from the thin, whip-like internal skeleton. It makes real sense to choose individuals with their polyps fully expanded during daylight.

WILL it reproduce in an aquarium?
There is little chance of this coral spawning in the home aquarium. However, as further inroads are made in the development of plankton substitutes, better-kept specimens may have every chance of producing gametes (sperm and eggs).

▼ *A low price and stunning good looks mean that this gorgonian is always in demand, but the money will be wasted unless you feed it regularly.*

Armoured crawlers

▶ Crustaceans belong to the largest grouping of animals – Phylum Arthropoda – which currently accounts for over 80% of the earth's known species. Of the 42,000 crustacean species described, most are marine in origin and many are available to hobbyists. Many more may arrive by accident in association with live rock or coral base rock. Crustaceans have a hard, protective exoskeleton that they must shed periodically in order to grow. The moulted material can closely resemble the original animal, often leading new aquarists to believe that some disaster has befallen their livestock. The problem is compounded because the newly moulted animal often becomes very elusive, hiding away in rockwork as its new exoskeleton hardens. The exoskeleton's rigidity also endows the animal with its mobility, which is compromised immediately post-moult.

Price guide

★	£2 – 10
★★	£10 – 20
★★★	£20 – 30
★★★★	£30 – 40
★★★★★	£40 – 50

PROFILE

A beautiful species of hermit crab, with alternating neon blue and black bands on its legs. Despite growing to a relatively large size for a dwarf species, it is largely benign and makes an excellent and useful addition to most reef aquariums, provided it is stocked with caution.

WHAT size?
Maximum of 6cm across the legs.

WHAT does it eat?
Algae, detritus and uneaten food intended for fish. Feed regularly when naturally occurring resources in the aquarium (algae) become used up.

WHERE is it from?
Australia and the Marshall Islands.

WHAT does it cost?
★☆☆☆☆
Price depends on size.

SIMILAR SPECIES

Could be confused with *Clibanarius tricolor,* which also has blue legs, but *C. tricolor* lacks the neon appearance of *C. elegans.*

Calcinus elegans

Blue knuckle hermit crab

HOW many in one tank?
A stocking rate of one for every 5-25 litres is recommended.

HOW compatible with other invertebrates?
Will kill and consume snails, such as *Astrea* spp. or *Turbo* spp., and then move into the shell. Providing spare shells and supplemental feeding can greatly reduce the chances of this occurring. Has also been reported removing the chitinous tubes from around fanworms.

WILL fish pose a threat?
Some fish are known to target hermit crabs, once they have worked out how to remove it from its protective shell. Dwarf angelfish have been observed doing this. All hermit crabs are vulnerable when changing shells.

WILL it threaten fish?
No. Too small and peaceful to present any threat to fish.

WHAT to watch out for?
Will usually mix nicely and peacefully with other species of hermit crab, particularly

▼ *Aquarists may forgive the clumsy nature of this crab in the light of its highly attractive appearance and usefulness in the aquarium.*

Paguristes cadenati. Choose specimens that are active and well acclimatised. Specimens have a good chance of achieving their maximum size in modest-sized aquariums and can be quite clumsy around poorly anchored invertebrates, knocking them off rockwork or turning them over as they forage.

WILL it reproduce in an aquarium?
Spawning can and does occur, but the planktonic larvae do not survive.

Calcinus laevimanus

Lefthanded dwarf hermit crab

This dwarf hermit crab is commonly imported together with more recognisable species from the Indo-Pacific. It is readily identified by its large, well-developed left pincer and the chocolate-brown walking legs with creamy yellow bands just above their tips. Fully grown specimens have a very robust appearance.

WHAT size?
Maximum 6cm across the legs.

WHAT does it eat?
Practically anything that it can get its pincers on! It will scavenge uneaten food intended for fish, plus algae and detritus.

WHERE is it from?
Widespread throughout the Indo-Pacific.

WHAT does it cost?
★☆☆☆☆
Inexpensive. Price depends on the size of the individual.

▶ *The blue bands running around the eye stalks distinguish this species from the very similar* Calcinus tibicen, *which hails from the Caribbean.*

HOW many in one tank?
Competition for shells can preclude the keeping of more than one specimen in aquariums below 250 litres that are not supplied with additional shells. Given sufficient space and shells, it is possible to keep a number of these crabs.

HOW compatible with other invertebrates?
Very efficient at killing and eating herbivorous snails. The large pincer is particularly adept at removing the snail's protective operculum. The crab then consumes its victim and may or may not move into its shell. Corals and sessile invertebrates should be safe, but *C. laevimanus* may kill and eat other hermit crabs. It will not usually directly harm sessile invertebrates, but once fully grown can overturn small specimens.

WILL fish pose a threat?
Vulnerable to fish that have developed techniques for removing hermits from their shells, particularly small ones. However, fully grown crabs should not be at risk from any reef-compatible fish.

WILL it threaten fish?
This opportunistic crab may kill and eat small, slow-moving species, but is often unfairly blamed for killing already ailing fish in the aquarium.

WHAT to watch out for?
When attractive and small (in shells measuring 10mm or less), this species will often tempt aquarists with little knowledge of its rapid growth rate. It is extremely hardy and useful in a reef aquarium, provided the aquarist remains aware of its relatively large final size and cosmopolitan tastes.

WILL it reproduce in an aquarium?
Yes. Breeding is not commonly experienced, but possible where numbers of this species are kept.

SIMILAR SPECIES
Several other species of crab are sold as 'dwarf', but many can prove at least as problematic as *Calcinus laevimanus*, including the Hawaiian zebra hermit, *Calcinus seurati,* and *Calcinus tibicen,* particularly where predation of molluscs is concerned.

Ciliopagurus strigatus

Cone shell hermit crab

PROFILE

A small and innocuous species that gives a great deal of pleasure to aquarists on account of its stunning good looks and benign behaviour.

WHAT size?
Maximum 6cm across the legs.

WHAT does it eat?
Primarily herbivorous, but reported to attack and eat snails for their shells. However, this species of hermit crab is adapted for life in the shells of the potentially dangerous cone snail, so perhaps this is not a bad thing in a home aquarium. Live rock-based systems are recommended, as they will provide plenty of natural food for this crustacean.

WHERE is it from?
Tropical Indo-Pacific. Particularly common from Hawaiian waters.

WHAT does it cost?
★☆☆☆☆ ★★☆☆☆
Prices depend on the size of the individual.

SIMILAR SPECIES
The large hermit species *Aniculus strigatus* has pronounced hairy legs and can be very destructive in a reef aquarium. However, it is rarely imported.

HOW many in one tank?
It is possible to keep a number of individuals in the same aquarium without too many issues, but provide plenty of spare shells in order to reduce competition for this resource.

HOW compatible with other invertebrates?
In the vast majority of aquariums this is a peaceful species, but it may attack stressed or dying animals.

WILL fish pose a threat?
Few, if any, reef-compatible fish are capable of threatening this hermit crab.

WILL it threaten fish?
No. They will scavenge dead fish, but are never responsible for the mortality.

▲ Aquarists may have to seek out specialist shell suppliers to provide increasingly larger homes for this attractive hermit crab.

WHAT to watch out for?
This species has a severely flattened body, an adaptation to the cone shells that other hermit crabs, with their rounded bodies, cannot exploit. Try to source some cone snail shells of varying sizes to facilitate the growth of the animal.

WILL it reproduce in an aquarium?
Sporadic reports of spawning events exist, but such experiences are highly likely where a number of individuals are housed together.

Clibanarius tricolor

Blue-legged dwarf hermit crab

PROFILE

This dwarf hermit crab is often stocked into a reef aquarium for its herbivorous characteristics, despite its penchant for eating arguably more useful algal grazers, such as *Astrea* and *Turbo* snails. It is more active during the day than many other dwarf hermit species.

WHAT size?
Maximum leg span 2.5cm.

WHAT does it eat?
Scours rock surfaces for almost any type of food, including algae and small animals. A few individuals in a live rock-based aquarium do not require supplementary feeding, but will seek out uneaten fish food and can prove a nuisance when you are attempting to target-feed sessile invertebrates.

WHERE is it from?
Caribbean Sea.

WHAT does it cost?
★☆☆☆☆
Inexpensive. Prices depend on the size of the individual.

HOW many in one tank?
A stocking rate of one for every 25 litres is acceptable, but most aquarists introduce far fewer. Introducing more individuals can result in a lack of food and therefore a high probability that this species will turn nasty!

HOW compatible with other invertebrates?
Will kill and consume snails such as *Astrea* spp. and then move into the empty shell. Can greatly reduce the overall diversity of life in a reef aquarium by removing small animals from the rock surfaces.

SIMILAR SPECIES
Several other species of crab are sold as 'dwarf', but many can prove at least as problematic as *Clibanarius tricolor*, including *C. seurati* and *Calcinus tibicen*, particularly where predation of molluscs is concerned.

WILL fish pose a threat?
Some fish are known to target hermit crabs once they have worked out how to remove them from their protective shells. All hermit crabs are vulnerable when changing shells.

WILL it threaten fish?
Too small to threaten most fish, but will happily kill and eat anything that comes within reach.

WHAT to watch out for?
Although recommended for the control of detritus and algae in a reef aquarium, the less active *Calcinus cadenati* is a better choice for most aquarists as it has a less voracious appetite.

WILL it reproduce in an aquarium?
Yes. Breeding is common when this species is housed in groups.

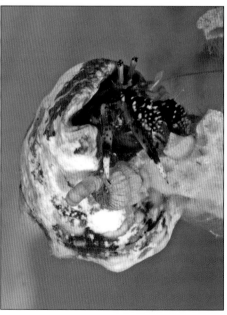

◀ *Where blue leg hermits are the sole herbivorous invertebrates stocked they can be highly useful, but do not mix them with other hermits or herbivorous snails.*

Dardanus megistos

White-spotted hermit crab

A large, potentially destructive hermit crab that is best kept in a species aquarium or in a system where its voracious appetite and clumsiness will not cause any damage.

WHAT size?
Maximum 13cm across the legs.

WHAT does it eat?
Primarily a scavenging animal that will accept a variety of different meaty foods in the aquarium. Offer dried algae in aquariums where natural green foodstuffs are not present. Will also predate small fish and sessile invertebrates.

WHERE is it from?
Tropical Indo-Pacific and Eastern Atlantic.

WHAT does it cost?
★★☆☆☆
Prices depend on the size of the individual.

▲ *The right-sized shell provides a haven for a white-spotted hermit.*

▲ *This crab is highly attractive yet entirely unsuitable for reef aquariums. Spare shells can also prove difficult to acquire, but can be sourced from specialist suppliers.*

HOW many in one tank?
Usually one.

HOW compatible with other invertebrates?
Will attack and eat tubeworms and molluscs. Very clumsy and highly mobile, so will knock over unattached invertebrates.

WILL fish pose a threat?
This hermit crab is large and tough enough to handle any reef compatible fish.

WILL it threaten fish?
Yes. Any fish small or weak enough to be caught will be consumed.

WHAT to watch out for?
Large shells can be difficult to

SIMILAR SPECIES
The genus *Dardanus* has many species available in the marine aquarium, including *D. pedunculatus*, which carries around anemones from the genus *Calliactis* on its shell. They afford the crab some protection through their stinging tentacles and benefit from the rather messy feeding behaviour of their host.

obtain. Given sufficient food, the growth rate of this species is prodigious.

WILL it reproduce in an aquarium?
Potentially, but few aquarists ever house more than a single specimen.

Manucomplanus varians

Antler crab

PROFILE

This species of hermit crab is a 'two-for-one' offering, as the 'shell' it is found in is actually a living, stinging hydroid colony called *Janaria mirabilis*, which secretes a stony skeleton. This offers the crab some protection against predators. The crab uses its enlarged, flattened pincer to seal the entrance to its home when it is in residence.

WHAT size?
Maximum of 4cm across the legs.

WHAT does it eat?
Algae, detritus and uneaten food intended for fish and invertebrates. It will scavenge almost anything in the aquarium, but it is worthwhile offering it supplemental feedings of chopped shellfish. Offering fine microplankton and phytoplankton may help to keep the *Janaria* healthy.

WHERE is it from?
Tropical Western Atlantic.

WHAT does it cost?
★★☆☆☆   ★★★☆☆
For a hermit crab this is an expensive purchase.

▶ *Check that both the coral and the crab are in good condition before making a purchase. Avoid specimens with algal growth or white patches on the 'shell'.*

HOW many in one tank?
Its high price means that it is usually maintained singly. Having multiple specimens may lead to aggression for 'shells'.

HOW compatible with other invertebrates?
Can attack damaged or stressed sessile invertebrates but is unlikely ever to be the instigator of the problem. May attack herbivorous snails for their shells, but such behaviour is rare.

WILL fish pose a threat?
The stinging cells of the hydrozoan host and its spiky growth form offer the crab some protection against the unwanted attentions of fish, but wrasse and some dwarf angelfish can attack it. Being able to seal its home against attack makes the crab fairly resistant to predation.

WILL it threaten fish?
No. It will scavenge dead fish but is never responsible for the mortality.

WHAT to watch out for?
Only buy specimens in which both animals appear healthy. Avoid specimens with signs of algal growth on the *Janaria* host. Some authorities maintain that the hermit crab will only survive when *Janaria* is healthy, but others have observed the crab moving into another shell and abandoning the dead host.

WILL it reproduce in an aquarium?
Although it may be possible for this species to mate and produce eggs in the aquarium this has not been reported, perhaps due to the lack of aquariums housing more than one individual.

Pagurites cadenati

Scarlet leg hermit crab

PROFILE

Probably the least problematic dwarf species of hermit crab for the reef aquarium. It is primarily nocturnal and will not always be on display, but it has a useful role in scavenging uneaten food intended for fish and will help to control nuisance algae.

WHAT size?
Maximum 4cm across the legs.

WHAT does it eat?
Scavenges uneaten fish food and picks algae from rocks.

WHERE is it from?
Caribbean Sea.

WHAT does it cost?
★☆☆☆☆
Not expensive, but many individuals are needed to fulfil a beneficial role in the reef aquarium.

SIMILAR SPECIES
Pagurites dagueti is collected from the Pacific coast of the United States. It is very similar to the Caribbean species, but grows to about ~~twice~~ the size and has white bands on the antennae, absent in *C. cadenati*.

▲ *A favourite with reef aquarists due to its modest proportions and beautiful appearance. Provide plenty of spare shells to avoid bickering.*

HOW many in one tank?
A stocking rate of one for every 5-25 litres is recommended.

HOW compatible with other invertebrates?
Can attack damaged or stressed sessile invertebrates. It is also reported to kill and eat grazing snails, such as *Astrea* spp., which leave behind a shell that can be suitable to move into.

WILL fish pose a threat?
Some fish appear to learn how to remove these crustaceans from their protective shells. Dwarf angelfish *(Centropyge* spp.) are particularly adept at this.

WILL it threaten fish?
No. They will scavenge dead fish but are never responsible for the mortality.

WHAT to watch out for?
Be sure to provide sufficient new shells for the crabs to move into. This can help to protect beneficial snails that might otherwise provide both a home and a meal for the crab.

WILL it reproduce in an aquarium?
Yes. Aquarists report frequent spawning events. Eggs are brooded inside the shell until the larvae emerge.

Calappa spp.

Shamefaced crab

Named for the stout pincers that lock beneath the front edge of the carapace, apparently reminiscent of a person hiding their face in shame, these crabs are occasionally imported for the aquarium trade. Worthy of sole occupancy of an aquarium in their own right, they require sand that is deep enough for them to burrow into.

WHAT size?
Most specimens do not exceed 10cm across the carapace.

WHAT does it eat?
In the wild, this group of crabs specialises in consuming hard-shelled prey, primarily molluscs. In an aquarium situation, they will accept most meaty foods.

WHERE is it from?
Circumtropical.

WHAT does it cost?
★★☆☆☆　★★★☆☆
Medium priced.

HOW many in one tank?
Usually kept singly.

HOW compatible with other invertebrates?
Its historically specialised tastes become more cosmopolitan with time in the aquarium and if allowed, it may consume corals, other crustaceans or sessile invertebrates.

WILL fish pose a threat?
This crab is a potential candidate for keeping in a live rock-based fish-only system with large fish in residence. However, it should not

SIMILAR SPECIES
Several species of shamefaced crab exist and some are imported for home marine aquariums. Perhaps the only relatively large species of true crab that finds its way into the hobby is the calico crab *Hepatus epheliticus*, for which the husbandry is similar.

▲ *The pincers lock into place in front of the carapace, protecting the mouthparts.*

be kept with triggerfish or pufferfish or any species that may like hard-shelled prey.

WILL it threaten fish?
Has the potential to catch and kill small or slow-moving fish on its nocturnal forays. However, it may be wrongly blamed for killing fish when actually scavenging carcasses.

WHAT to watch out for?
This fascinating crab has too many demands for most systems and should not be stocked without researching its requirements. Some aquarists are tempted to stock it into sumps or refugiums, but it is not suited to this existence. Its burrowing tendencies can undermine rockwork.

WILL it reproduce in an aquarium?
Unlikely to be kept in numbers, but may mate. Sexes are separate.

Camposcia retusa

Decorator crab

PROFILE

Although initially acquired because of its bizarre appearance, the decorator crab's welcome to the aquarium is often shortlived. Shortly after introduction, specimens take cuttings of prized corals in order to disguise themselves in keeping with their new environment. Small spines and hooks hold the adornments in place.

WHAT size?
Grows to at least 20cm across the span of the legs.

WHAT does it eat?
Consumes most meaty foods, such as chopped fish, shellfish and brineshrimp, mysis and flaked foods intended for fish.

WHERE is it from?
Tropical Indo-Pacific.

WHAT does it cost?
★★☆☆☆   ★★★☆☆
Medium priced.

SIMILAR SPECIES
There are several species of crab that use their environment to provide adornments. Some of the more delicate, smaller species require special attention, as they can be attacked and killed by fish or other crustaceans.

▲ *If you object to this crab pruning your corals then do not buy it. Such behaviour is inevitable. Similarly, buying a specimen because of the attractive growth present in the dealer's tank is pointless, as this will be shed when it changes its 'outfit' in your aquarium.*

HOW many in one tank?
Best kept singly.

HOW compatible with other invertebrates?
May harm the animals it prunes to provide its disguise, although the adornments to the legs and carapace will often survive. The good news is that if local conditions are suitable, these will continue to grow where they are dropped. This behaviour can thus help to spread sessile invertebrates around the aquarium, although it may result in specimen corals looking a bit untidy.

WILL fish pose a threat?
Few fish will harass this species

of spider crab. Indeed, its slow movement and decorations may mean that its tankmates do not even notice it.

WILL it threaten fish?
Unlikely. This species is more likely to avoid fish than try to attack them, but it may take advantage of sick fish or feed on corpses.

WHAT to watch out for?
Regardless of the diversity of life forms on its body when you buy this crab (commonly including sponges), it will almost certainly remove these and begin redecoration, often at the expense of colonial polyps such as zoanthids. Of course, when the crab moults in order to grow, it must find new decorations.

WILL it reproduce in an aquarium?
Not reported in captivity. Rarely, females brooding eggs may be imported.

Lybia tesselata

Boxer crab

PROFILE

This small and exquisitely beautiful species of true crab is particularly interesting because of its relationship with anemones. The latter spend much of their existence grasped between the crab's tiny pincers and are waved in the direction of a potential threat in a manner reminiscent of a pugilist shadow-boxing.

WHAT size?
10-15mm across the carapace.

WHAT does it eat?
This scavenging species is so well able to glean sufficient food from a live rock-based aquarium that it will rarely venture into the open, even when meaty foods are added to the aquarium. However, brineshrimp, mysis and chopped shellfish are useful foods, especially in a nanoreef aquarium with limited natural resources.

WHERE is it from?
Western Indian Ocean to the Marshall Islands in the Pacific Ocean.

WHAT does it cost?
★★☆☆☆
Relatively inexpensive

▲ *The larger aquarium will only afford occasional glimpses of the boxer crab.*

HOW many in one tank?
Can be kept singly or in groups.

HOW compatible with other invertebrates?
Will not harm other invertebrates.

WILL fish pose a threat?
Reef-compatible fish will ignore most specimens. Some large wrasse may attack the crab, but the anemones are a potent defensive adaptation and will deter many overly inquisitive fish.

WILL it threaten fish?
No. This crab will not harm even the smallest fish in a home aquarium.

WHAT to watch out for?
Although often elusive, this crab is worth keeping for the slightest possibility that you might witness it moulting. It will place its anemones safely nearby and shed its exoskeleton as quickly as possible before recovering its protectors and moving on.

WILL it reproduce in an aquarium?
Egg-carrying females are commonly imported, but this species will reproduce in the aquarium. The larvae are planktonic and therefore present many problems for the aquarist attempting to raise them.

SIMILAR SPECIES

There are other species of boxer crab that are very similar to *L. tesselata* and with very similar aquarium requirements. Crabs that have lost one of their anemones will often reappear with a pair again, which might suggest they play a role in encouraging the stinging animal to reproduce asexually.

Mithraculus sculptus

Emerald crab

PROFILE

A popular species of crab, closely related to the spider crabs *(Majidae)*. This primarily herbivorous species is often sold as a potential grazer of unwanted forms of algae, such as *Valonia* and *Ventricaria,* the so-called 'bubble algae'.

WHAT size?
3-4cm across the carapace.

WHAT does it eat?
Like most true crabs, the emerald is opportunistic. It will graze algae, scavenges in the wild and accepts most offerings in an aquarium.

WHERE is it from?
Caribbean, Western Atlantic.

WHAT does it cost?
★★☆☆☆

HOW many in one tank?
Can be maintained on their own or in groups. Maximum numbers will be determined by food availability; they can misbehave if there is an insufficient supply.

HOW compatible with other invertebrates?
Reported to be entirely reef-safe, but being an opportunist, it may attack injured, stressed or dying specimens.

WILL fish pose a threat?
Most reef-safe fish will leave this crab alone. It is vulnerable to larger wrasse species, especially when its exoskeleton is soft following a moult.

WILL it threaten fish?
May attack sick fish but this is rare, even for the largest crabs. The vast majority of specimens are completely benign.

SIMILAR SPECIES
M. forceps is a brown species from the same region as *M. sculptus,* and with identical characteristics. Some very closely related species, referred to as 'Mithrax' crabs, are imported on coral base rock or with live rock. They can be identified by the spoon-shaped ends to their pincers and lumpy-looking carapace. Although rogues do turn up, the vast majority of the hitchhikers are completely harmless, even when approaching their maximum size.

WHAT to watch out for?
Introducing specimens to control a bloom of unwanted algae is rarely successful, but these crabs are interesting to care for in their own right. They may occasionally eat calcareous forms of algae in a home aquarium.

WILL it reproduce in an aquarium?
Possibly, where the crabs are kept in groups. Female crabs have wider tails than males as an adaptation for holding the egg mass, once laid.

◀ *A small, attractive crab that can play a role in the control of certain forms of algae, in particular those avoided by other herbivores.*

Neopetrolisthes ohshimai

Anemone crab

PROFILE

Anemone crabs are not true crabs. Instead, they are closely related to the squat lobsters and porcelain crabs, as evidenced by their 'flapping' tail used in swimming, primarily to avoid predators. They often live on or beneath anemones, where present, hence their common name.

WHAT size?
2-3cm across the carapace.

WHAT does it eat?
Anemone crabs have a pair of specialised feeding appendages used to capture particulate food from the water column. Although they will capture meaty items, such as brineshrimp, they also benefit from the addition of zooplankton and similar particulate offerings.

WHERE is it from?
Tropical Indo-Pacific.

WHAT does it cost?
★★☆☆☆
Relatively inexpensive.

▶ *The feeding appendages of the anemone crab are held open in prevailing currents and capture fine particulate food. These are then drawn through the mandibles, which remove the trapped morsels from the netlike bristles.*

HOW many in one tank?
Can be kept singly or in a small group where space and host anemones allow.

HOW compatible with other invertebrates?
Will not harm other invertebrates. Completely reef-safe.

WILL fish pose a threat?
Most fish will ignore this crab, especially if it is resident in an anemone. Large wrasse are a possible exception. It can be driven off by resident anemonefish if it attempts to share their anemone. In common with most crustaceans, it is vulnerable after a moult.

WILL it threaten fish?
No. This species is more likely to avoid fish than try to attack them.

WHAT to watch out for?
Introduce specimens with care. They have the disconcerting habit of dropping their pincers when roughly handled. Do not attempt to drop them directly into a host anemone, particularly carpet anemones (*Stichodactyla* spp.), as they might be consumed by the potential host!

WILL it reproduce in an aquarium?
Possibly, but not reported in captive specimens.

Percnon gibbesi
Sally lightfoot crab

PROFILE

This species of crab is often sold as a useful scavenger and herbivore for the reef aquarium, fulfilling this role for months before becoming a nuisance and potentially very destructive. Its flattened body enables it to hug the rocky substrates it calls home in areas of strong surge.

WHAT size?
Grows to at least 10cm across the span of the legs.

WHAT does it eat?
Almost anything, but mostly meaty foods and algae. As it grows and matures, its tastes become more cosmopolitan and it begins to threaten other animals in the aquarium.

WHERE is it from?
Circumtropical.

WHAT does it cost?
★☆☆☆☆   ★★☆☆☆
Relatively inexpensive.

SIMILAR SPECIES
Few species of crab resemble the Sally lightfoot.

◄ *The flattened carapace and legs of the Sally lightfoot crab indicate its preference for life in areas with strong currents and in surge zones. It is best suited to a live rock-based aquarium with large, non-predatory fish, where it will be a useful scavenger.*

HOW many in one tank?
Can be maintained in groups but often kept singly.

HOW compatible with other invertebrates?
May harm hermit crabs, small ornamental crabs, anemones and other invertebrates in the aquarium. Smaller specimens are usually relatively trustworthy, but their disposition changes as they grow.

WILL fish pose a threat?
Very few fish will attack this species of crab. Those that might are seldom, if ever, maintained in reef aquariums.

WILL it threaten fish?
Yes. Larger specimens are known to catch and kill fish. As the crab is very difficult to locate and remove once stocked, only house it with fish much larger than itself.

WHAT to watch out for?
Healthy specimens will be active and react quickly to sudden movements inside and outside the aquarium. Choose small specimens and keep them well fed in an attempt to minimise future problems.

WILL it reproduce in an aquarium?
This is possible in captivity but rarely reported. Very occasionally, females brooding eggs may be imported.

Pilumnus spp.

Teddy bear crab

PROFILE

The teddy bear crab is named for its beige, hairy appearance and is an enigmatic creature. The vast majority of specimens will be imported by accident in live rock or coral base rock. Aquarium behaviour depends very much on the individual; some are benign, while others prove extremely problematic.

WHAT size?
10cm across the carapace.

WHAT does it eat?
All specimens will scavenge uneaten food intended for fish or other invertebrates. Some have more cosmopolitan tastes and will actively predate sessile invertebrates.

WHERE is it from?
Tropical Indo-Pacific.

WHAT does it cost?
☆☆☆☆☆
Unlikely to be offered for sale. Usually free.

SIMILAR SPECIES
The genus *Pilumnus* contains many species of 'hairy crab', but few others ever make their way into marine aquariums.

HOW many in one tank?
Not usually kept intentionally, but specimens that behave themselves appear not to be territorial.

HOW compatible with other invertebrates?
Some specimens will behave impeccably, even fulfilling a beneficial role by scavenging uneaten food and detritus. Others eat corals, shrimp, hermit crabs and snails.

WILL fish pose a threat?
Fish are unlikely to threaten these crabs, which are very shy under most circumstances. They may be vulnerable when they moult.

WILL it threaten fish?
Potentially, large specimens can catch and eat small, slow-moving species of fish.

WHAT to watch out for?
The presence of this crab may only be detected following the aftermath left when it predates a coral. It will seldom emerge during daylight and even large specimens might go unnoticed by the aquarist.

WILL it reproduce in an aquarium?
Possible where more than one individual is present, but perhaps not to be encouraged. The larvae are extremely unlikely to survive in the home aquarium.

▼ *It is difficult to generalise about this almost unmistakable species, due to the different characteristics of individuals. The impact it has on your aquarium may depend entirely on the specific corals present and the personality of the animal.*

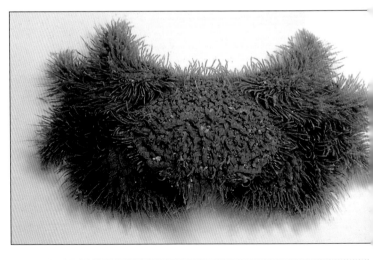

Stenorhynchus seticornis

Arrow crab

A small-bodied species with a large leg span. Historically, it was used to control populations of free-living polychaete worms, when the latter were all deemed to be undesirable in the reef aquarium. The arrow crab is a less popular choice today.

WHAT size?
A 4cm carapace could mean a 30cm leg span!

WHAT does it eat?
Wild specimens are scavengers. They will accept a variety of meaty, shellfish-based foods, including flake, in the home aquarium.

WHERE is it from?
Caribbean.

WHAT does it cost?
★★☆☆☆   ★★★☆☆

SIMILAR SPECIES
There are numerous species from tropical seas around the world. The decorator crab (see page 101) is closely related, but its body is adorned with a number of spikes for attaching pieces from the animal's surroundings. It is far less predatory than the arrow crab in a home aquarium.

HOW many in one tank?
Best maintained singly.

HOW compatible with other invertebrates?
Will eat a variety of other invertebrates, given the opportunity. Apart from feeding on polychaete worms, it is known to 'winkle' dwarf hermit crabs out of their shells and seek out moulting ornamental shrimps. It will also consume small snails and anything else it can get its pincers on.

WILL fish pose a threat?
Most reef-compatible fish will ignore this species, but it is vulnerable when moulting and when small (10cm or so across the legs).

WILL it threaten fish?
Yes. Large specimens, in particular, may catch and eat slow-moving fish. Sometimes

▲ *The slender, almost fragile, appearance of the arrow crab belies its potential for causing problems in otherwise peaceful, small reef aquariums.*

blamed for killing fish that were already sick or on the verge of death.

WHAT to watch out for?
To avoid potential problems, only house the arrow crab with large, benign fish and sturdy corals. Small specimens often tempt aquarists with little knowledge of their large final size, achieved quickly even in modestly proportioned aquariums. These relative giants have appetites to match!

WILL it reproduce in an aquarium?
Possible, but given its aggressive nature with conspecifics, it can be difficult to maintain a pair.

Alpheus bellulus

Tiger pistol shrimp

PROFILE

The tiger pistol shrimp is the most commonly available of the pistol shrimp known to partner gobies. It burrows extensively and relentlessly. A pair of these shrimp with a mated pair of gobies makes a fascinating addition to any aquarium. Housing them in a smaller system will increase your chances of being able to watch them in action.

WHAT size?
6-8cm body length, excluding the antennae.

WHAT does it eat?
Offer brineshrimp, mysis and other shellfish-based diets. Place the food close to one of the burrow's entrances, as this shrimp is unlikely to venture too far from its refuge.

WHERE is it from?
Tropical Indo-Pacific. Most aquarium specimens are imported through Sri Lanka.

WHAT does it cost?
★★☆☆☆
Relatively inexpensive.

SIMILAR SPECIES
Alpheus rapax and *A. djeddensis* are similar in behaviour, if not appearance. They are less commonly encountered in dealers' tanks but are similarly benign.

HOW many in one tank?
Keep singly or in male-female pairs.

HOW compatible with other invertebrates?
Will not directly harm other invertebrates. However, its tunnelling behaviour has the potential to undermine poorly constructed reefs in the home aquarium.

WILL fish pose a threat?
In an aquarium where these shrimp are kept with 'watchman' gobies they are often safe from all but the most persistent predators. However, dottybacks can harass specimens persistently, but the burrowing nature of this shrimp means that it has a refuge during its most vulnerable period – immediately post moult.

WILL it threaten fish?
No.

WHAT to watch out for?
Because it produces loud noises

LIGHT SABRE
Other Alpheids that do not have a cooperative relationship with gobies can potentially attack small fish using the sound and, somewhat surprisingly, light emitted by the claw to stun their prey.

with its snapping pincer that sound like breaking glass, it is often confused with the mantis shrimp that is actually able to break glass! Tunnelling is undertaken almost without pause, so aquarists who enjoy a neat, sandy substrate might want to avoid this animal.

WILL it reproduce in an aquarium?
Yes. This is one of the easiest pistol shrimps to sex. Males have relatively larger snapping pincers than females. Pairs will usually occupy the same burrow, which they excavate themselves.

◀ *It is possible to identify this A. bellulus as a male specimen by its relatively large snapping claw. Its burrow is shared by a pair of Cryptocentrus cinctus gobies.*

Alpheus randalli

Randall's pistol shrimp

PROFILE

A beautiful shrimp that belongs to a larger group commonly known as pistol, or snapping, shrimps. One of its pincers is greatly enlarged to accommodate the musculature that forms the priming mechanism for the snapping claw. Many members of the genus *Alpheus* form commensal relationships with certain gobies, such as *Cryptocentrus* and *Amblyeleotris*.

WHAT size?
3-4cm body length, not including antennae.

WHAT does it eat?
Primarily a scavenging species that will feed on most meaty foods, such as brineshrimp, mysis and chopped shellfish. It should also accept flake and granular offerings.

WHERE is it from?
Western Pacific.

WHAT does it cost?
★★☆☆☆   ★★★☆☆
Medium priced.

▲ *A stunning pair.* A. randalli *is pictured here with the antenna goby* (Stonogobiops nematodes)*.*

HOW many in one tank?
Best kept singly or in male/female pairs.

HOW compatible with other invertebrates?
Will not directly harm other inverts. However, its tunnelling behaviour has the potential to undermine poorly constructed reefs in the home aquarium.

WILL fish pose a threat?
In aquariums where these shrimp are kept with 'watchman' gobies, they are often safe from all but the most persistent predators. Being so small, the shrimps are particularly vulnerable when first introduced and any fish might look upon them as a snack.

WILL it threaten fish?
This species will not harm fish, but this is not true of all alpheid pistol shrimps.

WHAT to watch out for?
Specimens are often available at small sizes that are easily predated or 'lost' in larger systems. Preparing a refuge for them can help. For example, you can push a 10cm-length of 10mm-diameter pipe into the substrate as a starting point for their excavations. Lowering their transport bag to the substrate in this area before releasing them can result in them remaining in an area where they will be easier to see.

WILL it reproduce in an aquarium?
Yes. Male pistol shrimps can be distinguished by their relatively larger snapping pincers. Pairs will usually occupy the same burrow, which they excavate themselves.

DETERRENT SNAP

The snapping claws of most commensal pistol shrimps are primarily used at a deterrent to predators and rivals from the same species, rather than as an offensive weapon. Always be sure of your identifications where pistol shrimps are concerned, as you do not want to introduce a free-living predatory species by accident!

Alpheus soror
Bullseye pistol shrimp

PROFILE

This is a beautiful pistol, or snapping, shrimp, with a more benign disposition than other free-living members of the genus *Alpheus*. It will readily occupy existing nooks and crannies in the aquarium and seems to prefer the boundary of sand or gravel and rock to make its home.

WHAT size?
Specimens achieve 8cm or so.

WHAT does it eat?
Wild specimens feed on a variety of small creatures and fish, and scavenge dead tissue. Aquarium specimens accept any meaty foods, such as shellfish flesh, brineshrimp and mysis.

WHERE is it from?
Indian Ocean and tropical Pacific. Most aquarium specimens are imported through Sri Lanka.

WHAT does it cost?
★★☆☆☆
★★★☆☆

▶ *The colours of this pistol shrimp species immediately endear it to aquarists. Although less likely to excavate tunnel systems than symbiotic species, it will nonetheless move gravel and small rocks.*

HOW many in one tank?
Could be housed in male-female pairs in the larger aquarium.

HOW compatible with other invertebrates?
May predate very small crustaceans and molluscs.

WILL fish pose a threat?
Any predatory species, including dottybacks (*Pseudochromis* spp.), pufferfish and lionfish pose a threat.

WILL it threaten fish?
May threaten very small fish such as gobies, but provided it is well fed any aggression is rare.

WHAT to watch out for?
Unlike other popular aquarium pistol shrimps, this is a free-living species that does not dig extensive burrows. It may be difficult to locate at feeding time, but once settled will become confident enough to venture into the open when it detects food in the water.

SIMILAR SPECIES
The bullseye pistol shrimp is difficult to confuse with other species of snapping shrimp, not least because of its magnificent coloration. Many other free-living forms of these shrimps are available in the hobby, albeit sporadically, but many are quite efficient predators that employ their snapping claw to stun potential prey items, such as small fish and other shrimps.

WILL it reproduce in an aquarium?
Not reported but highly likely. The fry will be difficult to raise without specialist materials, such as dedicated rearing tanks and live, planktonic foods. Male specimens possess larger 'snapping' pincers than females.

Gnathophyllum americanum

Bumblebee shrimp

PROFILE

A small but beautiful species of shrimp that makes a fascinating addition to the nanoreef or small species aquarium. Its natural diet suggests that it could be difficult to maintain in any system, but it acclimatises well and in the absence of predators makes a very interesting species to care for.

WHAT size?
2cm maximum.

WHAT does it eat?
Offer brineshrimp, mysis and finely chopped shellfish-based diets. The shrimp's natural diet is the tube feet of echinoderms, but these are not necessary for its survival.

WHERE is it from?
Tropical Indo-Pacific. The genus is circumtropical in distribution.

WHAT does it cost?
★★☆☆☆
Relatively inexpensive.

SIMILAR SPECIES

The genus *Gnathophyllum* contains a number of species and all feed on the tube feet of starfish. Husbandry requirements are the same for each.

HOW many in one tank?
House only one pair unless the aquarium is especially large, in which case other pairs can be added, but the chances of seeing them are reduced!

HOW compatible with other invertebrates?
Will eat the tube feet of starfish and sea urchins, but is otherwise reef-safe. House only one pair per aquarium due to their territorial nature.

WILL fish pose a threat?
Yes. Any species that will consume a shrimp as large as mysis will make short work of a bumblebee shrimp. This is why it is vital to house the shrimps in a small aquarium without any remotely predatory fish.

▲ A diminutive species that requires a small, specialised aquarium if it is to be safe from harm, yet remain visible on a regular basis.

WILL it threaten fish?
No. Far too small to present a threat to any fish species.

WHAT to watch out for?
Acclimatise this shrimp carefully over a number of hours.

WILL it reproduce in an aquarium?
Yes. This species can spawn. Pairs are territorial and easy to spot as the sexes will sit in close proximity to one another in the dealer's tank. To date, it has not been successfully reared in captivity.

Hymenocera picta

Harlequin, or clown, shrimp

PROFILE

Considering its very strict dietary requirements, this stunning crustacean is still commonly imported for the aquarium trade.

WHAT size?
Females 6cm body length, males 4cm body length, not including antennae. Antennae are very long – sometimes 60cm or more across.

WHAT does it eat?
This shrimp feeds exclusively on living starfish, so cannot be stocked into most aquariums, as it will consume ornamental *Linckia* and *Fromia* species with gusto.

WHERE is it from?
Tropical Indo-Pacific.

WHAT does it cost?
★★☆☆☆   ★★★☆☆
Medium priced.

SIMILAR SPECIES

The unique nature of these shrimps is underlined by the fact that the genus only contains one species. *Hymenocera elegans* is often used as the species name for this shrimp.

▲ *If not supplied with its correct, highly specific diet, the beautiful harlequin shrimp will not survive in the aquarium.*

HOW many in one tank?
Usually acquired in pairs. Keep one pair per aquarium.

HOW compatible with other invertebrates?
Other than the obvious threat to starfish posed by this specialist feeder, it is completely reef-safe. There are reports that it sometimes attacks the tube feet of other echinoderms, such as urchins, but this may be out of desperation rather than choice.

WILL fish pose a threat?
The usual suspects predate these shrimp, notably the otherwise reef-safe pseudochromid dottybacks and hawkfish.

WILL it threaten fish?
No. This is a harmless species of shrimp that, unusually, does not share the opportunistic tendencies of most crustaceans that find their way into the marine aquarium.

WHAT to watch out for?
Despite being a potentially difficult shrimp to care for (unless you are willing to provide plentiful live starfish), the harlequin shrimp can fulfil a useful role in the reef aquarium. Starfish from the genus *Asterina* can reproduce asexually in reef aquariums and some species have been observed attacking corals. In this circumstance the harlequin shrimp may offer a reasonable biological control on their numbers.

WILL it reproduce in an aquarium?
Yes. Pairs are commonly available. However, noticing when the shrimps are entering a spawning cycle can be difficult.

Leander plumosus

Long nose, or plumose, shrimp

PROFILE

This infrequently imported shrimp has an unusual appearance, with hairlike bristles on the rostrum, which also bears large serrations. It is generally found associated with macroalgae in shallow water. For this reason, it is likely to feel most at home in an aquarium with long leafy forms of algae, particularly the *Sargassum* spp. brown forms that commonly grow from live rock imported from Fiji and Indonesia.

WHAT size?
3-4cm body length.

WHAT does it eat?
A scavenger and predator of tiny invertebrates in the aquarium. Will accept most finely chopped or particulate diets, including brineshrimp, mysis and shellfish.

WHERE is it from?
Tropical Indo-Pacific.

WHAT does it cost?
★★☆☆☆
Relatively inexpensive.

▶ *The unusual plumose shrimp justifies inclusion in a peaceful reef aquarium by virtue of its unique yet attractive appearance.*

HOW many in one tank?
Usually kept singly, but it is tolerant of its own kind.

HOW compatible with other invertebrates?
Will not harm other invertebrates.

WILL fish pose a threat?
Will be attacked and eaten by hawkfish, dottybacks and any other species with a penchant for small crustaceans. Best maintained with peaceful gobies, dartfish and similar species. It does not clean

SIMILAR SPECIES

Many shrimps have a very obvious hump in their bodies, but none have the highly distinctive coloration of *L. plumosus*.

fish, so does not enjoy the relative invulnerability that this role endows on many shrimp species imported for the aquarium hobby.

WILL it threaten fish?
No. This shrimp is too delicately built to attack even the smallest fish.

WHAT to watch out for?
The pronounced hump in the body is entirely natural, as is the patch of pigmentation that tops its peak. Acclimatise it slowly and carefully, as it does not enjoy sudden changes in water chemistry.

WILL it reproduce in an aquarium?
Little is known about reproduction in this species, but it is highly likely to spawn in aquariums where more than one individual is present.

Lysmata amboinensis

Skunk cleaner shrimp

PROFILE

This perennial favourite with marine aquarists is widely available in the hobby and will do well in almost any reef aquarium.

WHAT size?

6cm body length, excluding antennae.

WHAT does it eat?

In its natural environment, this species obtains much of its food by removing parasites from fish. It will also enter the mouths of large predatory species and remove the dead tissue that remains there after a meal. In the aquarium, it accepts any meaty food, including shellfish, chopped fish and mysis.

WHERE is it from?

Tropical Indo-Pacific.

WHAT does it cost?

★★☆☆☆
Relatively inexpensive.

SIMILAR SPECIES

Lysmata grabhami is from the tropical Atlantic, but has a thin white stripe on each edge of the predominantly red tail. Husbandry is the same as for *L. amboinensis*.

▶ *The skunk cleaner becomes increasingly bold as it grows.*

HOW many in one tank?

Keep in multiples of two in a larger aquarium (450 litres or more). In smaller systems, two is adequate, ideally introduced simultaneously.

HOW compatible with other invertebrates?

May steal food intended for other sessile invertebrates. It can disturb species it chooses to rest on, but this is usually temporary.

WILL fish pose a threat?

Predatory species, such as *Pseudochromis,* can attack and kill even large shrimps, but most fish would rather take advantage of their cleaning services.

WILL it threaten fish?

Large specimens are opportunistic and may seize weak or very small fish. In most cases they do not present a problem to marine fish.

WHAT to watch out for?

It will often moult very soon after introduction to its new home, leaving behind a ghostlike facsimile of its exoskeleton. This often leads aquarists to believe that their specimen has died, but is perfectly normal behaviour. This species does well when introduced in pairs of similarly-sized individuals.

WILL it reproduce in an aquarium?

Yes. Spawning events are common. Small individuals are functionally male but become hermaphrodite with increasing size. The larvae have been successfully reared in captivity.

Lysmata debelius

Fire, or blood, cleaner shrimp

PROFILE

This deepwater species of 'cleaner' shrimp is perhaps the most desirable and expensive from the genus *Lysmata*. Although only discovered relatively recently, it has become a firm favourite with reef aquarists, despite its moderately disruptive characteristics.

WHAT size?
6cm body length, excluding antennae.

WHAT does it eat?
Wild specimens will obtain some nourishment by removing parasites from visitors to their cleaning station. In an aquarium, offer meaty foods including flake, mysis and coarsely chopped shellfish. The latter is particularly useful to keep these shrimp occupied while you feed sessile invertebrates.

WHERE is it from?
Tropical Indo-Pacific. Most specimens are sourced through Sri Lanka.

WHAT does it cost?
★★☆☆☆　★★★☆☆

▶ *Choose well-settled specimens of fire shrimp that have been fed regularly in the dealer's aquarium.*

HOW many in one tank?
Keep in multiples of two in a larger aquarium (450 litres or more). In smaller systems, two is adequate, ideally introduced simultaneously.

HOW compatible with other invertebrates?
May steal food intended for other sessile invertebrates. It can disturb species it chooses to rest on, but this is usually temporary. Reported to eat small polyps, but this may depend on the individual shrimp concerned.

WILL fish pose a threat?
Predatory species from the

SIMILAR SPECIES
Few aquarium species have the vivid coloration of the fire cleaner shrimp, but in many ways the husbandry of all members of the genus *Lysmata* is similar.

genus *Pseudochromis* can attack and kill even large shrimps, but most fish prefer to take advantage of their cleaning services or else ignore them completely.

WILL it threaten fish?
Large specimens are opportunistic and may seize weak or very small fish. In most cases they do not present a problem.

WHAT to watch out for?
Specimens can be reclusive, reflecting the fact that they are found in low-light conditions experienced at depth. It often helps to add a pair before introducing any fish. Repeated feeding will soon coax them into the open and they will continue to emerge when they are hungry.

WILL it reproduce in an aquarium?
Yes. Spawning events are common. Small individuals are functionally male, but become hermaphrodite with increasing size. The larvae have been successfully reared in captivity.

Lysmata wurdemanni

Peppermint shrimp

PROFILE

Although less colourful than some its close relatives in the genus *Lysmata,* the peppermint shrimp has achieved widespread popularity due to its role in the control of the nuisance anemone *Aiptasia*. However, its effectiveness will depend on many variables, including the availability of other food sources.

WHAT size?
4cm body length, excluding antennae.

WHAT does it eat?
Will accept most small meaty foods in an aquarium. Eats *Aiptasia* when easier-to-tackle foodstuffs are not available, but may also consume similar small polyps, purposely stocked by the aquarist.

WHERE is it from?
Caribbean.

WHAT does it cost?
★☆☆☆☆
★★☆☆☆
The price
depends
on size and
exact origin.

HOW many in one tank?
Can be kept in groups of six or more, even in quite modestly-sized aquariums (200 litres).

HOW compatible with other invertebrates?
Will predate small polyps, particularly yellow polyps, but this can be discouraged by providing alternative foods. It is common for peppermint shrimps to consume all available *Aiptasia* and leave similar polyps unharmed.

WILL fish pose a threat?
Many fish will not recognise this shrimp as a cleaner species and therefore attempt to eat it. They include anemonefish, hawkfish, pseudochromids and similar.

WILL it threaten fish?
No, but all shrimps from this genus are opportunistic and take advantage of a free or easy meal.

▼ *The aiptasia-eating peppermint shrimp is more reclusive than many other* Lysmata *spp. shrimp.*

WHAT to watch out for?
Tank-raised specimens are sporadically available and often very small, despite commanding a high price. Do not add them to an aquarium well stocked with fish. Indeed, any size of peppermint shrimp is best introduced to an aquarium devoid of fish livestock, as this way they will become more confident and prove more efficient at consuming *Aiptasia*.

WILL it reproduce in an aquarium?
Yes. The larvae of this species are some of the easiest of all ornamental shrimp species to raise in captivity.

SIMILAR SPECIES

Several species resemble *L. wurdemanni* very closely. Some are from temperate or subtropical waters and therefore unsuitable for a tropical marine aquarium. The dietary preferences of other species are largely unknown, making them potentially dangerous to reef inhabitants. One other species, *Rhynchocinetes uritai*, is known as the peppermint, or dancing, shrimp. It is highly attractive, but not safe in a mixed reef display. However, it is a rewarding species to keep in a live rock-based, fish-only system stocked with benign fish.

Odontodactylus scyllarus

Peacock mantis shrimp

PROFILE

Although commonly referred to as a 'shrimp', this crustacean belongs to a group called the Stomatopods and is only distantly related to the true shrimp. It is named for the paired raptorial appendages that it uses to strike or seize its prey. It is classified according to predation style. Those that feed on hard-shelled prey, such as crabs and molluscs, have clublike appendages and are called 'smashers'. Those that capture prey from the water column, particularly fish, are called 'spearers'.

WHAT size?
15cm, not including antennae. Large specimens are available within the hobby, but may have to be ordered from your dealer.

WHAT does it eat?
In its natural environment, this 'smasher' prefers hard-shelled prey, including crabs and molluscs, but will accept most meaty foods in the aquarium. It relishes shellfish, with or without shells, and similar meaty foods.

WHERE is it from?
Tropical Indo-Pacific.

WHAT does it cost?
★★☆☆☆   ★★★★☆
Price depends on size.

▲ O. scyllarus *is an awesome animal for the stomatopod enthusiast.*

HOW many in one tank?
Keep singly.

HOW compatible with other invertebrates?
Do not keep with any hard-shelled crustaceans and molluscs. Indeed, best kept in a species aquarium on its own.

WILL fish pose a threat?
Given the opportunity, dottybacks and hawkfish will consume small mantis shrimps.

WILL it threaten fish?
Potentially, yes. In theory, 'smashers' should not show any interest in fish as prey items, but a large specimen will use its raptorial appendages as defensive tools and most aquarium fish will suffer as a result of a punch from one of these shrimp.

WHAT to watch out for?
Mantis shrimp have a reputation for being able to break glass up to 6mm thick, so must be kept in a suitably strong aquarium. They are also capable of breaking the fingers of a careless aquarist. Use a feeding prong to feed this shrimp and always establish exactly where it is lurking before putting your hands in the aquarium to carry out essential maintenance.

WILL it reproduce in an aquarium?
Yes, but as both male and female are highly territorial, there are inherent problems in providing an aquarium suitable for reproduction to take place. If you wish to attempt to raise some mantis shrimp larvae (which has been achieved in captivity) it would be best to look for a female already carrying a clutch of eggs.

SIMILAR SPECIES

Many species of mantis shrimp can be obtained by marine aquarists. Some arrive by accident in live rock or associated with coral base rock. Ideally, they should be removed alive and rehomed. Some species remain small, and provided they are kept well fed, seldom prove a problem. Larger species can wreak havoc if left in situ.

Periclimenes brevicarpalis

Clown anemone shrimp

PROFILE

The genus *Periclimenes* embraces a large number of small shrimps that live on a variety of host organisms. *P. brevicarpalis* is most commonly found living on the tentacled surface of carpet anemones *(Stychodactyla spp.)* or pizza anemones *(Cryptodendrum adhaesivum).*

WHAT size?
Females are larger, reaching around 3cm; males 5-10mm smaller.

WHAT does it eat?
Clown anemone shrimps will remove food particles from the surface of their host and sometimes feed on their faecal material. Offer finely chopped shellfish, chopped mysis and brineshrimp in an aquarium. May also benefit from the addition of zooplankton-based products. Target-feeding may be necessary, as the shrimp will be reluctant to stray far from its host to feed.

WHERE is it from?
Tropical Indo-Pacific.

WHAT does it cost?
★★☆☆☆
Relatively inexpensive.

▶ *Clown anemone shrimp are known to live for extended periods in the absence of their usual host and thus make suitable candidates for a peaceful nanoreef aquarium.*

HOW many in one tank?
Best kept in male/female pairs, one pair to an aquarium.

HOW compatible with other invertebrates?
Will not harm invertebrates, with the possible exception of its host.

WILL fish pose a threat?
Dottybacks and hawkfish together with some wrasse species present the greatest threat, even if the shrimp is housed in a host anemone with a potent sting.

WILL it threaten fish?
No. Harmless towards all fish.

WHAT to watch out for?
There are reports of some species of anemone shrimp possibly feeding on the tentacles of their host. Although this is not necessarily the case for *P. brevicarpalis,* take care to provide enough food for the shrimp to help discourage this behaviour.

SIMILAR SPECIES
A large number of species within the genus *Periclimenes* could be imported for the aquarium trade. It is not worth considering any of these unless you know the specific host(s) for a particular species of shrimp and can provide an environment suitable for both. For example, *P. imperator* is known to use both sea cucumbers and nudibranchs as host species.

WILL it reproduce in an aquarium?
Yes. Females are commonly imported in a gravid state. Where pairs are housed in a suitable aquarium, reproduction is highly likely.

Rhynchocinetes durbanensis

Dancing shrimp

PROFILE

Sometimes confusingly referred to as the 'peppermint shrimp', this species of small tropical shrimp is lovely to look at but inconsistent in the control of the nuisance anemone, *Aiptasia* spp. Indeed, it can be highly problematic when introduced into an aquarium in the mistaken belief that it will adopt a beneficial role. It is best suited to a live rock-based, fish-only aquarium that is home to peaceful species.

WHAT size?
4cm body length.

WHAT does it eat?
Primarily a scavenger and opportunistic feeder on polyps and small corals. It will readily accept most meaty foods, including shellfish, mysis and krill.

WHERE is it from?
Tropical Indian Ocean and Western Pacific.

WHAT does it cost?

★☆☆☆☆
★★☆☆☆
Inexpensive.

HOW many in one tank?
Can be kept in groups, but males will sometimes fight. Usually, one male to two or three females is acceptable.

HOW compatible with other invertebrates?
Eats colonial polyps, such as zoanthids and star polyps (*Briareum, Clavularia* and *Pachyclavularia*). May also target other species of sessile invertebrate.

WILL fish pose a threat?
Small specimens will be attacked and eaten by hawkfish, dottybacks and any other species with a penchant for crustaceans. Larger individuals are more resistant to unwanted attentions from fish.

WILL it threaten fish?
May snatch small fish such as gobies, but this is rare. It will scavenge injured, sick or dead fish and is sometimes

SIMILAR SPECIES
The genus *Rhynchocinetes* contains a number of species that are very similar in appearance. *R. durbanensis* is the commonest in the hobby, exported through Sri Lankan sources, together with aquarium favourites, such as the cleaner shrimp (*L. grabhami*) and the fire shrimp (*L. debelius*).

wrongly blamed for fish deaths caused by poor conditions or husbandry.

WHAT to watch out for?
As with all shrimps, acclimatise this species slowly. Specimens should be alert and active, especially when food is added to the aquarium. The shrimp's common name comes from its slightly nervous sideways movement when standing in the open.

WILL it reproduce in an aquarium?
Mating and spawning is to be expected. Males are often slightly larger and have larger, more robust pincers than the females.

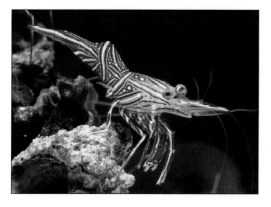

In the right system, and in the absence of the polyps that it is known to consume, the dancing shrimp is a useful scavenger.

Saron marmoratus

Marbled shrimp

A commonly imported shrimp, but not suitable for most reef aquariums. However, it is an ideal candidate for a live rock-based, fish-only system that is home to peaceful fish. Here, it can do little harm and you can enjoy its good looks.

WHAT size?
7-8cm body length.

WHAT does it eat?
Primarily a scavenger and opportunistic feeder on polyps and small corals. Readily accepts most meaty foods, including shellfish, mysis and krill.

WHERE is it from?
Tropical Indo-Pacific.

WHAT does it cost?
★★☆☆☆
Relatively inexpensive.

SIMILAR SPECIES
A few shrimps from the genus *Saron* are available in the hobby. The purple-clawed *Saron rectirostris* is particularly beautiful. Other species show marked sexual dimorphism, with males having very long pincers.

▲ *Saron shrimp are beautiful, but cannot be stocked in coral-rich systems.*

HOW many in one tank?
Can be kept singly, in male-female pairs or small groups.

HOW compatible with other invertebrates?
Eats colonial polyps such as zoanthids. May also target other species of sessile invertebrate.

WILL fish pose a threat?
Small specimens will be attacked and eaten by hawkfish, dottybacks and any other species with a penchant for crustaceans. Larger individuals are more resistant to unwanted attentions from fish.

WILL it threaten fish?
May snatch small fish such as gobies, but this is rare. They will scavenge injured, sick or dead fish and are sometimes wrongly blamed for fish deaths caused by poor conditions or husbandry.

WHAT to watch out for?
As with all shrimps, acclimatise this species slowly. Specimens should be alert and active when food is added to the aquarium.

WILL it reproduce in an aquarium?
Mating and spawning are to be expected. Males are often more robust in appearance and have longer pincers than the females.

Stenopus hispidus

Boxing shrimp

PROFILE

This is one of the largest shrimps commonly imported into the aquarium hobby. It is named for the large pincers that it uses in territorial defence. In its natural environment, this crustacean cleans large fish species. In the aquarium, few boxing shrimps will be housed with fish large enough to benefit from their cleaning capabilities.

WHAT size?
Females 6cm, males 4cm body length, excluding the antennae. The antennae are very long, sometimes measuring 30cm or more.

WHAT does it eat?
Accepts most small meaty foods.

WHERE is it from?
Tropical Indo-Pacific.

WHAT does it cost?
★★☆☆☆
Price depends on size and origin. Pairs may command a higher premium.

▶ *Beware of buying temptingly small individuals of this shrimp. They grow rapidly and the larger they become, the more potential harm they can cause.*

HOW many in one tank?
Keep singly or in male/female pairs. (True pairs are commonly available.)

HOW compatible with other invertebrates?
In common with most crustaceans found in marine aquariums, this is an opportunistic animal that may predate small hermit crabs or other shrimp, given the opportunity.

WILL fish pose a threat?
Small individuals are under threat from hawkfish and pseudochromids. Large specimens can be vulnerable when they undergo a moult.

WILL it threaten fish?
Potentially, yes. However, as it will scavenge corpses, it is often wrongly blamed for predatory behaviour. It may relentlessly harass sleeping or slow-moving fish.

WHAT to watch out for?
This shrimp grows potentially very large and can prove a nuisance to fish and corals.

WILL it reproduce in an aquarium?
Yes. It is possible to buy pairs and they will often spawn, but given the slightly elusive nature of this species in a home aquarium, the event will often go unnoticed by the aquarist.

SIMILAR SPECIES
Several *Stenopus* species are available in the hobby, but you may have to wait for them to arrive with your dealer. They include *Stenopus zanzibaricus, S. cyanoscelis, S. tenuirostris* and *S. pyrsonotus*. The first three species are all smaller than *S. hispidus*.

Thor amboinensis

Sexy shrimp

PROFILE

A small but beautiful species of commensal shrimp that shares its existence with stinging corals and anemones. Its common name is derived from the wiggling of its abdomen while at rest.

WHAT size?
2cm body length, excluding antennae.

WHAT does it eat?
Will accept most small meaty foods. Not afraid to remove the meals of its host or, indeed, feed upon their waste.

WHERE is it from?
Ubiquitous in tropical reefs.

WHAT does it cost?
★☆☆☆☆ ★★☆☆☆
Price depends on size and origin.

SIMILAR SPECIES

The genus *Thor* has a number of species, but few others are ever imported for the aquarium hobby.

▶ *With its wiggling rump, the aptly named sexy shrimp is beautiful and immediately endearing.*

HOW many in one tank?
Keep in groups of three or more individuals whenever possible.

HOW compatible with other invertebrates?
May steal food intended for host sessile invertebrates. It has also been suggested that it could feed directly on its host, but there are few records of actual damage being caused.

WILL fish pose a threat?
Most fish that are capable will have a go at eating this shrimp, including anemonefish, hawkfish, pseudochromids and similar.

WILL it threaten fish?
No. This species is too small to offer any threat to reef fish.

WHAT to watch out for?
The sexy shrimp is sometimes accidentally imported with stony corals, such as *Euphyllia* spp.

WILL it reproduce in an aquarium?
Yes. Spawning events are common where groups are housed together, but due to the small size of the individuals concerned, only rarely reported.

Urocaridella antonbruunii

Cave cleaner shrimp

PROFILE

A wonderful shrimp that is not frequently seen at present, but its popularity should increase as more aquarists become familiar with what it has to offer.

WHAT size?
3-4cm body length, excluding antennae.

WHAT does it eat?
Wild specimens clean other fish, removing parasites and dead skin. In the aquarium they readily accept most small meaty foods, such as brineshrimp, chopped mysis and similar.

WHERE is it from?
Tropical Indo-Pacific.

WHAT does it cost?
★★☆☆☆
Relatively inexpensive.

SIMILAR SPECIES

There are more *Urocaridella* species yet to be described by science. Most have the same basic transparent appearance, but differ in the colour and pattern of the spots located on the body. The husbandry for each appears to be the same as for *U. antonbruunii*.

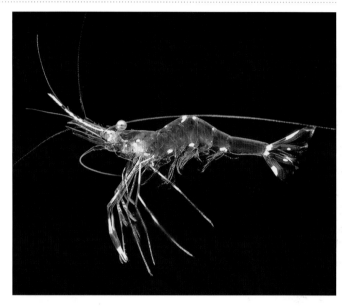

HOW many in one tank?
Can be kept singly, in pairs or small groups.

HOW compatible with other invertebrates?
Will not harm other invertebrates.

WILL fish pose a threat?
This shrimp is known to clean very large species of fish, such as moray eels and groupers, so the main threat to its safety will come from smaller predatory fish, such as hawkfish and dottybacks.

WILL it threaten fish?
No, it is too delicately built to attack even the smallest fish.

▲ *A beautiful and surprisingly hardy shrimp in the absence of predators.*

WHAT to watch out for?
The cave cleaner shrimp often uses the legs located on the underside of its abdomen for propulsion and is often seen swimming in open water. Be sure to protect circulation pumps located inside the aquarium, otherwise this shrimp can easily be sucked into them.

WILL it reproduce in an aquarium?
There is little information available, but it is highly likely to spawn in aquariums where more than one individual is present.

Allogalathea elegans

Crinoid squat lobster

These diminutive crustaceans are not commonly available, but determined aquarists with the patience to seek them out can find them. *A. elegans* remains small throughout its life and is one of a number of species that uses a variety of echinoderms, particularly feather starfish (crinoids), as hosts. Of these the black-and-white species is the most commonly encountered.

WHAT size?
Grows to about 1cm total body length, excluding pincers.

WHAT does it eat?
Wild specimens almost certainly pick detritus and food particles from the feathery arms of their host, thereby providing a service, albeit at the expense of a proportion of the echinoderm's food. In the home aquarium they accept most small meaty particulate material up to the size of, and including, brineshrimp.

WHERE is it from?
Tropical Indo-Pacific.

WHAT does it cost?
★☆☆☆☆   ★★☆☆☆
Inexpensive.

▶ *At rest on a brown-coloured feather starfish this crustacean is immediately noticeable, yet on a black host it is practically invisible.*

HOW many in one tank?
Can be maintained singly or in small numbers.

HOW compatible with other invertebrates?
Will not harm other invertebrates. At risk from shrimps and crabs that may feed on this tiny crustacean.

WILL fish pose a threat?
Almost any fish is likely to make

a meal of small squat lobsters. Their attractive coloration helps to disguise them when ensconced in the arms of their host, but not when crawling over rocks in a reef aquarium. Best maintained in a peaceful nanoreef aquarium in the absence of predatory fish.

WILL it threaten fish?
No. Harmless towards fish.

WHAT to watch out for?
Squat lobsters are closely related to porcelain crabs, such as *Neopetrolisthes* (see page 104). They have the habit of shedding their pincers when threatened, and although this is regrettable it is perfectly natural. Healthy specimens will regrow their pincers as they moult.

WILL it reproduce in an aquarium?
Not reported but possible.

Enoplometopus daumi

Pink reef lobster

PROFILE

These 'dwarf' lobsters are beautiful in appearance if not in temperament. They have received a mixed press over the years and it is likely that their impact on a reef aquarium will depend largely on the particular mix of animals in the aquarium. Opinions on their aquarium compatibility are highly polarised. This may be a reflection on the temperaments of individual specimens, rather than a generalisation about each species.

WHAT size?
Grows to at least 10cm total body length, excluding pincers.

WHAT does it eat?
This is an omnivorous species that readily accepts most meaty foods, including shellfish, mysis, krill and similar diets. It will also eat macroalgae and may benefit from the addition of dried algae to the aquarium.

WHERE is it from?
Tropical Indo-Pacific.

WHAT does it cost?
★★☆☆☆   ★★★☆☆
Medium priced.

Small reef lobsters are tempting, but reclusive and often destructive.

HOW many in one tank?
Maintain singly or in male-female pairs.

HOW compatible with other invertebrates?
Given the opportunity, it may kill and eat hermit crabs, crabs and shrimps. All these potential prey items are particularly vulnerable immediately after the moult when their bodies are still soft, which makes rapid movement difficult. Molluscs are also at risk, but the lobster does not generally harm sessile invertebrates.

WILL fish pose a threat?
Very few fish considered reef-compatible will bother any of the species of reef lobsters commonly available in the aquarium trade.

WILL it threaten fish?
Yes. Has been reported stalking and killing a variety of fish in the aquarium, particularly those that seek refuge in the rockwork at night.

WHAT to watch out for?
Small specimens are very appealing, but do not buy them on impulse. They make very interesting additions to an aquarium where they can do little harm, for example one with a mixture of colonial polyps, plus their own company and no other.

WILL it reproduce in an aquarium?
It is possible to house male and female specimens in the same aquarium. The sexes can be determined by examining the genital structures located between the legs on the underside of the animals. They have spawned in captivity, with the male copulating with the female immediately after she moults.

SIMILAR SPECIES

At least two other species are sold as reef lobsters and both are members of the genus *Enoplometopus*. *E. occidentalis* has an orange-red body, punctuated with white spots. *E. debelius* is covered in red-pink spots and has an attractive pale pink coloration with yellow tips to the legs and pincers.

Panulirus versicolor

Blue spiny lobster

Spiny lobsters, sometimes referred to as crayfish, lack the large pincers of reef lobsters from the genus *Enoplometopus*. Their large robust antennae, believed to have a role in food capture, are a key to their identification. In the blue spiny lobster these antennae are white.

WHAT size?
Grows to around 40cm body length. The antennae can be over twice this length.

WHAT does it eat?
A carnivorous crustacean that enjoys chopped shellfish, mysis, brineshrimp and almost any meaty food.

WHERE is it from?
Tropical Indo-Pacific.

WHAT does it cost?
★★☆☆☆   ★★★☆☆
Medium priced.

SIMILAR SPECIES
Other spiny lobsters imported for the aquarium trade are very few and far between, but they are available. Most will have the potential to grow large and they should not be acquired without research into their long-term requirements.

HOW many in one tank?
Best maintained singly, unless housed in an exceptionally large aquarium, where pairs can be safely accommodated.

HOW compatible with other invertebrates?
Likely to kill and eat a variety of animals in the marine aquarium, including other crustaceans, gastropod molluscs and many soft-bodied animals. This tendency for predating tankmates increases with age.

WILL fish pose a threat?
No fish generally considered to be reef-safe presents a threat to the spiny lobster.

WILL it threaten fish?
Yes. Will kill and eat fish that prefer to use refuges in rockwork during the night.

WHAT to watch out for?
Although it is not suited to the type of reef aquarium commonly maintained by aquarists, given the right mix of sessile invertebrates and large robust fish, this lobster makes a decent aquarium animal that should prove hardy and long-lived. Never be tempted to impulse buy it, even though it is extremely tempting when small.

WILL it reproduce in an aquarium?
Such events are rare, as most aquarists will shun mature specimens, which are too large to be easily housed. However, reproduction cannot be ruled out in a large aquarium.

▼ *In the right system spiny lobsters make interesting inhabitants, but they are not suitable for the average mixed reef aquarium.*

Scyllarus spp.

Slipper, or locust, lobster

PROFILE

Slipper lobsters are bizarre-looking crustaceans in which the antennae have evolved as flattened structures that meet at the centre of the carapace to form a ploughlike structure used in digging. The genus *Scyllarus* contains species that stay relatively small, but others are occasionally available that require a larger aquarium for their long-term husbandry.

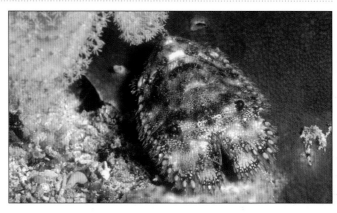

WHAT size?
Members of the genus *Scyllarus* reach 6-10cm, which makes them more suitable for the majority of reef aquariums than their larger relatives. These may achieve a substantial 40cm or more.

WHAT does it eat?
Slipper lobsters are carnivores that feed on a variety of soft-bodied invertebrates, such as worms and molluscs. They are known to kill and eat some bivalve molluscs and will probably remove useful detritivores, such as polychaete bristleworms, from sandy substrates.

WHERE is it from?
Circumtropical.

WHAT does it cost?
★★☆☆☆ ★★★☆☆
Medium priced.

HOW many in one tank?
Best kept singly.

HOW compatible with other invertebrates?
Will not usually harm sessile invertebrates, such as hard and soft corals or colonial polyps. May attack *Tridacna* spp. clams, especially if they are located on the sand.

WILL fish pose a threat?
Few fish will harass slipper lobsters, although these are particularly vulnerable after moulting.

WILL it threaten fish?
Not recorded attacking fish and not really equipped to do them any harm.

WHAT to watch out for?
If a specimen is offered for sale, it is vital to research the species in order to establish its maximum size. Species that grow large are in the majority

▲ *Slipper lobsters are interesting but reclusive animals, with a characteristic body shape.*

SIMILAR SPECIES
Members of the genera *Scyllarides*, *Arctides* and *Thenus* are sometimes available, but most of these grow in excess of 30cm long. They are therefore likely to cause problems with digging and moving rocks as they make their way around the aquarium.

when it comes to aquarium importation. Such animals need a large aquarium and are potentially destructive.

WILL it reproduce in an aquarium?
Not reported, but possible where male and female specimens are housed together.

Shelled gliders

▶ The molluscs, represented by perhaps 100,000 or more species, show a large degree of diversity. Generalising about their body design is difficult, since all of the major groups show varying degrees of specialisation. However, the basic plan centres around the mantle – a body wall that surrounds the internal organs – a muscular foot and a shell secreted from calcium carbonate. The evolutionary process has finely tuned these components into the highly diverse species we recognise as snails, slugs and bivalve shellfish. In some species the shell has become lost or reduced. Others retain an internal shell, invisible to the naked eye. The Phylum Mollusca contains not only highly useful species for marine aquarists, but also some very problematical hitch-hikers, stocked accidentally when they arrive in association with live rock or coral base rock.

Price guide

★	£1 – 5
★★	£5 – 15
★★★	£15 – 30
★★★★	£30 – 75
★★★★★	£75-100

PROFILE

This small group of mainly tropical herbivorous snails is commonly stocked in aquariums to help control unwanted forms of algae. It is frequently, but erroneously, referred to as the 'turbo' snail.

WHAT size?
Large specimens may reach 4-5cm shell height.

WHAT does it eat?
It is particularly fond of diatoms, but will enjoy most film-forming microalgae, providing they are not toxic, such as cyanobacteria and some dinoflagellates.

WHERE is it from?
Tropical Caribbean and Western Atlantic.

WHAT does it cost?
★☆☆☆☆
Price depends on size and source.

SIMILAR SPECIES

The genus *Astraea* contains many species with similar grazing habits and modest size. Other genera, such as *Trochus*, *Lithopoma* and *Tectus*, contain similar species in terms of appearance and habits.

Astraea spp.

Star snail

HOW many in one tank?

The stocking rate of this snail will depend on the power of your lighting, the amount of other livestock present, particularly herbivores, and the overall nutrient load of the system.

HOW compatible with other invertebrates?

Not really large enough to cause a problem by knocking over other invertebrates. Exclusively herbivorous, so will not harm other invertebrates.

WILL fish pose a threat?

Opportunistic fish may take advantage of specimens that have fallen from the rockwork

▲ *An iconic species of herbivorous snail that is worth every penny and more of its modest price.*

or glass and are unable to right themselves (see later). Most fish ignore them.

WILL it threaten fish?

No. This species presents no danger towards any fish.

WHAT to watch out for?

Acclimatise new specimens slowly. Place them by hand onto rockwork (not sand), with their opening downwards. Astraea snails, together with a number of other herbivorous species, are prone to falling from

rockwork and being unable to right themselves, particularly if the spire of their shell falls into a rock crevice. Check regularly for upended specimens.

WILL it reproduce in an aquarium?

Yes. This small snail is frequently observed spawning, but the planktonic larvae have not been raised in the home aquarium.

▼ *Stock Astraea spp. with other, larger species, such as* Turbo *or* Trochus, *as they complement each others' grazing preferences.*

Cerithium spp.

Cerith

PROFILE

Readily distinguished from other snails regularly offered for sale by the long spire on the shell. There are several species of cerith suitable for the reef aquarium. They are primarily herbivorous and fulfil a useful role in the aquarium, not only on account of their grazing, but also because they keep the sand turned over and aerated.

WHAT size?
Size depends on the species concerned, but few of the commonly imported ones exceed 3cm shell length.

WHAT does it eat?
In addition to the various types of microalgae, particularly diatoms, found in the reef aquarium, this snail will also consume small amounts of detritus. In many species this is ingested as they swallow sand and thus they may also extract other material from the sand. Other species graze from hard surfaces in much the same way as turbo or trochus snails (see page 137).

WHERE is it from?
Circumtropical.

WHAT does it cost?
★☆☆☆☆ ★★☆☆☆
Inexpensive.

HOW many in one tank?
Usually maintained in numbers to take full advantage of their beneficial role in the aquarium. Gradually increase numbers according to the quantity of available natural food.

HOW compatible with other invertebrates?
Unlikely to bother any invertebrates directly. Always at risk of predation from crabs and hermit crabs. The latter will often move into the vacant shell once they have enjoyed their meal.

WILL fish pose a threat?
Very few fish considered to be reef-compatible will bother cerith snails.

WILL it threaten fish?
This snail presents no threat to any fish species.

WHAT to watch out for?
Many imported individuals take some time to recover from the rigours of transportation and

▲ Ceriths are characterised by the tall spires of the shell and its pale colour, which closely resembles the colour of the substrate in which the animals live.

appear inactive. Avoid these where possible and try to choose specimens that are crawling over the aquarium glass or rockwork.

WILL it reproduce in an aquarium?
Spawning events are commonplace and usually occur at night. The egg masses are usually visible as large, squiggly trails across aquarium glass and rocks. Unfortunately, cerith larvae are planktonic and few, if any, ever survive.

ADDED BONUS
Some authorities report that this is one of the few animals that will feed on slime algae (cyanobacteria), but this will depend entirely on the species concerned.

SIMILAR SPECIES
Cerith snails could be confused with dove snails or mud snails. There are also some species of carnivorous mitre snails with shell spires equal in size to that of the ceriths.

Chiton spp., *Acanthopleura* spp., *Stenoplax* spp.

Chiton

PROFILE

These primitive marine snails are readily identified by the multiple plates that make up their shell. Most of the specimens encountered by aquarists will be accidental imports on live and coral base rock, but larger species are also offered for sale.

WHAT size?
In some species, shell length can reach 30cm, but most are much smaller.

WHAT does it eat?
Consumes algal films, including diatoms. Take care when introducing specimens where other herbivorous snails are present, as this animal is almost impossible to feed once the aquarium's natural supplies of algae have been exhausted.

WHERE is it from?
Circumtropical. Also encountered in temperate and subtropical waters.

WHAT does it cost?
★★☆☆☆
Price depends on size and origin.

HOW many in one tank?
Usually maintained singly.

HOW compatible with other invertebrates?
Some species have been reported to be opportunistic and to consume a small amount of tissue from corals, but such instances are very rare. Usually completely reef-safe.

WILL fish pose a threat?
Few fish are able to tackle chitons. They are vulnerable when overturned, but their strong muscular feet seldom let go of the substrate voluntarily.

WILL it threaten fish?
No.

WHAT to watch out for?
Specimens should have very strong attachment to the substrate. Acclimatise them over a prolonged period, as they

SIMILAR SPECIES
The commonest species of chiton encountered tend to be slug chitons from the genus *Cryptoplax,* which are elongate and wormlike, with greatly reduced protective plates. They are frequently imported with live rock from Indonesia.

are sensitive to sudden changes in salinity and temperature.

WILL it reproduce in an aquarium?
Some smaller species can reproduce prolifically, but larger specimens offered for sale seldom spawn.

▼ *The commonest chitons offered for sale show the shieldlike arrangement of plates along their dorsal surface that is typical of the group.*

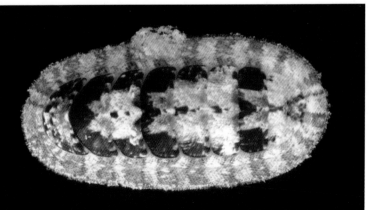

Haliotis spp.

Abalone

PROFILE

An unusual snail that is more widely known for the mother-of-pearl lining to its 'mule's ear' shell and its value as a food item. These molluscs are useful in the control of algae, but their demand for plentiful supplies can be their downfall, as they rapidly consume it and then starve, unless the aquarist intervenes.

WHAT size?
May reach 15cm shell length.

WHAT does it eat?
Feeds exclusively on algae. It particularly enjoys diatom films, but will also consume certain forms of filamentous algae. Offering dried nori can help specimens that have consumed all other algae in the aquarium.

WHERE is it from?
Tropical Indo-Pacific.

WHAT does it cost?
★★☆☆☆ ★★★☆☆
Price depends on size and origin.

HOW many in one tank?
A single specimen will consume a large amount of algae. Do not introduce this mollusc in numbers unless you can guarantee sufficient food in the long term.

HOW compatible with other invertebrates?
The only harm that this snail may inflict on invertebrates is indirect, perhaps by dislodging unanchored corals as it browses on the rockwork.

SIMILAR SPECIES
May resemble large specimens of the hitch-hiking snail *Stomatella varians*. Otherwise, abalone are very distinctive animals; their curved shells are punctuated by a row of holes that increase in diameter towards the outer margin.

WILL fish pose a threat?
Some fish may nip at the mollusc's lateral tentacles, but given that it is more likely to be nocturnal in its habits, it can avoid significant harassment by retreating into rockwork during the day.

WILL it threaten fish?
No.

WHAT to watch out for?
Abalone must be acclimatised over a prolonged period as they are sensitive to sudden water changes. Provided they are active in the dealer's aquarium and adhere strongly to the substrate, it is safe to buy them quite soon after importation, as they will probably be starving through lack of suitable food.

WILL it reproduce in an aquarium?
Unlikely, although not impossible.

▶ Abalone are secretive herbivores with cryptic coloration. They are readily identifiable by the series of holes along the outer edge of the shell, through which water flows after it has passed over the gills.

Hydatina spp.

Bubble snail

PROFILE

Bubble snails are beautiful animals, but infrequently imported. Do not buy them unless you are sure of the natural diet of the species concerned. They are more closely related to sea slugs than other marine snails and cannot retract their mantle completely inside the shell.

WHAT size?
The mantle extends beyond the shell. Can reach 4-5cm shell length and over 10cm total length.

WHAT does it eat?
Hydatina species tend to feed on polychaete 'bristleworms' that may be found in large numbers in a mature reef aquarium. Other bubble snails consume diatoms and other film algae.

WHERE is it from?
Circumtropical. Also encountered in temperate and subtropical waters.

WHAT does it cost?
★★☆☆☆   ★★★☆☆
Price depends on size and origin.

◀ *The gorgeous mantle of the bubble snail extends far beyond the shell and is as delicate as it appears. Don't be tempted by the snail's good looks unless you are able to provide its long-term needs.*

HOW many in one tank?
Usually kept singly, but can be maintained in numbers, providing there is sufficient food.

HOW compatible with other invertebrates?
May present a threat to beneficial worms that form natural populations in the marine aquarium. Otherwise harmless or even potentially useful. Should not be housed with large crabs or hermit crabs.

WILL fish pose a threat?
Fish may nibble at the mantle of this species, but this is likely to be out of inquisitiveness, rather than any desire to feed on the snail.

WILL it threaten fish?
No.

WHAT to watch out for?
Providing the correct diet for this species is essential. Research a potential purchase before buying it. Acclimatise slowly. The large, delicate mantle means that these snails do not do well in an aquarium with vigorous currents.

WILL it reproduce in an aquarium?
Possible but unlikely. Will depend to a large extent on the provision of plentiful rations.

SIMILAR SPECIES
Genera offered for sale, however sporadically, include *Bulla* and *Haminoea*. These tend to feed on film algae that occur naturally in the reef aquarium.

Nassarius spp.

Nasser, or mud, snail

PROFILE

Several species known as nassarius, or nasser, snails are currently available in the hobby, albeit sporadically. They are highly useful animals, but as might be expected, some are more suitable for most reef aquariums than others. Many keep the sand turned over and aerated as they move through it.

WHAT size?
Size will depend on the species involved, but few of the commonly imported ones will reach more than 2cm shell length.

WHAT does it eat?
These snails primarily scavenge dead material, but happily consume uneaten food intended for fish. Offering small chunks of shellfish is also a good idea.

WHERE is it from?
Circumtropical.

WHAT does it cost?
★☆☆☆☆ ★★☆☆☆
Inexpensive.

▶ *Mud snails are wonderful creatures that can fulfil a highly useful role in the reef aquarium, consuming detritus and uneaten food intended for fish.*

HOW many in one tank?
Can be maintained in large numbers, providing there is sufficient food.

HOW compatible with other invertebrates?
Should not harm any of their tankmates, although it is sensible to establish the natural diet of the species you intend to buy.

WILL fish pose a threat?
Few fish will bother nassarius snails, but several species of wrasse commonly kept in the reef aquarium may crunch very small snail specimens. Stock genera such as *Pseudocheilinus* and *Macropharyngodon* with caution.

WILL it threaten fish?
This snail presents no threat to any fish species.

WHAT to watch out for?
Individuals should be active and respond quickly to the introduction of food to the aquarium. They are hardy souls, but it is wise to acclimatise

them over an extended period. Individuals spawned and grown in an aquarium environment make long-lived and sturdy specimens.

WILL it reproduce in an aquarium?
Yes. Many species will spawn, depositing their eggs in capsules that are often visible on the aquarium glass. At least one species has direct larval development that results in a self-sustaining population that expands, provided that there is sufficient food to support them.

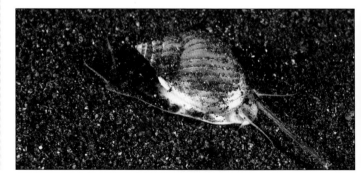

Nerita spp.

Nerite snails

PROFILE

These small herbivorous snails are found in tropical regions and can have a role in controlling undesirable algae in the reef aquarium. Several species are offered for sale, but only a few are suitable for inclusion in the average reef system.

WHAT size?
Reaches about 2-3cm.

WHAT does it eat?
Algal films. These snails particularly enjoy diatoms.

WHERE is it from?
Circumtropical. Many specimens imported for the aquarium are collected from the Caribbean and Hawaii.

WHAT does it cost?
★☆☆☆☆
Inexpensive.

SIMILAR SPECIES

Several genera may be imported, including *Puperita* and *Neritina*. Their suitability for the marine aquarium will depend on their behaviour. Those that spend all their time below water are the most desirable.

HOW many in one tank?
Can be maintained singly or in numbers, as long as there is a plentiful supply of natural food to meet the long-term requirements of several specimens.

HOW compatible with other invertebrates?
Larger species of crab may occasionally prise this snail from the rockwork to eat it. Nerites will not harm any invertebrates.

WILL fish pose a threat?
Fish do not generally bother nerite snails.

WILL it threaten fish?
No.

WHAT to watch out for?
Many species of nerite are collected from intertidal zones,

▲ *Nerites are small, inconspicuous snails that appear to be equally at home out of water as in it, so watch your step!*

where they spend extended periods out of the water. This behaviour is mirrored in the aquarium; the snail can overheat and die when exposed under intense lighting and may even leave the aquarium altogether. Such species are really only suitable to a specialised tidal aquarium.

WILL it reproduce in an aquarium?
Nerite snails lay eggs in a capsule similar to that of the dove snail *(Euplica* spp.), often on aquarium glass. Unlike dove snail larvae, those of nerites are planktonic and do not survive.

Pusiostoma mendicaria

Bumblebee snail

PROFILE

A small but beautiful species of marine gastropod that is prized by many aquarists for its good looks and the beneficial roles it performs. Its omnivorous diet and ability to keep sand beds aerated means that it is always in demand.

WHAT size?
Seldom grows to more than 1cm shell length.

WHAT does it eat?
Scavenges dead material, such as uneaten food intended for fish, grazes algae and consumes detritus. It is seldom necessary to offer this snail species any supplemental feeding, as few of its preferred foodstuffs are in short supply in most aquariums.

WHERE is it from?
Pacific Ocean.

WHAT does it cost?
★☆☆☆☆
Inexpensive.

HOW many in one tank?
Can be kept singly or in numbers.

HOW compatible with other invertebrates?
Has been reported consuming polyps, particularly zoanthids. Otherwise largely harmless. Small hermit crabs may kill the snail and use its shell.

WILL fish pose a threat?
Avoid species with a taste for hard-shelled animals. Of the fish species generally considered to be reef-safe, wrasse probably represent the greatest threat to this mollusc.

WILL it threaten fish?
Presents no threat to any fish.

WHAT to watch out for?
Try to choose active specimens, rather than those resting on the substrate. If individuals are crawling over glass or rockwork, this is usually an indication that they are in good health.

SIMILAR SPECIES
The bumblebee snail is similar in overall shell shape to the dove snails (*Euplica* spp.), a herbivorous gastropod that often arrives as a hitch-hiker on live rock. Dove snails achieve a similar size; a specimen measuring 10mm is viewed as fully gown. They should be welcomed into the aquarium.

WILL it reproduce in an aquarium?
Spawning has not been recorded for this species. It may be stimulated by environmental cues that do not occur in most aquariums. However, given the bumblebee snail's small size and the fact that it is usually kept in numbers, the possibility of courtship cannot be ruled out.

▶ *The number and width of the vertical gold bands surrounding the shell of the bumblebee snail are highly variable and may not be present at all in this species.*

Turbo spp.

Turban, or spiral, turbo snail

PROFILE

This almost legendary grazer of algae is a very popular addition to most reef aquariums. It has a large appetite for a variety of algae, including filamentous forms. Its favourite meal is probably brown diatomaceous algae.

WHAT size?
Large specimens reach 6cm or so.

WHAT does it eat?
Various types of algae growing on rock and glass in the aquarium. Will not usually accept dried forms of macroalgae.

WHERE is it from?
Tropical Indo-Pacific.

WHAT does it cost?
★☆☆☆☆   ★★☆☆☆
Price depends on size and source.

▼ *Trochus* spp. *can achieve equally substantial proportions as* Turbo *spp. and are consequently just as clumsy in the aquarium.*

▲ *The turbo snail is named for its whorled shell. It will consume a wider range of algae than* Astraea *spp., including filamentous forms.*

HOW many in one tank?
Keep singly or in groups. Do not be tempted to overstock these snails, despite their useful role in consuming algae. Build up numbers according to the abundance of their natural food.

HOW compatible with other invertebrates?
Does not present any direct threat, but is sufficiently large and powerful to push over corals and cause rockfalls as it moves around the aquarium. Shrimp and hermit crabs may predate it. The most significant threat is the rare predatory polychaete worm *Oenene fulgida*, which suffocates the snail with mucus secreted about the shell opening.

WILL fish pose a threat?
Few fish present a threat to this robust snail. However, large anemonefish have been

SIMILAR SPECIES
The giant turbo snail *Trochus* sp. requires much the same care and consumes similar types of algae. With regard to the control of filamentous ('hair') algae, both species are better in a preventative role, but where present in sufficient numbers they are capable of eliminating it from an afflicted aquarium. Both species require a slow process of acclimatisation before being introduced into a new aquarium.

observed knocking it into their host anemones or flipping it over, which can often be fatal for this mollusc.

WILL it threaten fish?
No. This species presents no dangers towards any fish.

WHAT to watch out for?
Always place newly acquired specimens onto rocks the right way up. They have limited self-righting abilities and will die or be eaten if left in any other position. They will not eat cyanobacteria (slime algae) and should not be placed on top of blooms of this unsightly growth.

WILL it reproduce in an aquarium?
Yes. This broadcast-spawning snail commonly spawns in the aquarium, but the larvae are planktonic.

Aplysia spp.

Sea hare

PROFILE

The sea hare is so named for the large sensory tentacles on its head. This substantial species of sea slug has an internal shell that is not visible externally. The only thing the sea hare eats is algae, so it is commonly sold as a solution to problems with nuisance algae. However, it has a large appetite and can quickly consume all the suitable food in an aquarium. Aquarists must consider what they will do with a specimen once it has fulfilled its purpose. It should be returned to a dealer rather than allowed to starve.

WHAT size?
May reach 15cm body length.

WHAT does it eat?
Feeds exclusively on algae. Small specimens will consume filamentous forms and some will browse on higher forms of macroalgae.

WHERE is it from?
Tropical Atlantic-Caribbean; Tropical Indo-Pacific.

WHAT does it cost?
★★☆☆☆   ★★★☆☆
Price depends on size and origin.

⬆ *Sea hares are large nudibranchs with an anti-predation response that involves expelling ink in a similar way to cephalopod molluscs. Their skin is also loaded with toxins.*

HOW many in one tank?
Best kept singly unless housed in a very large aquarium with an abundance of their natural food.

HOW compatible with other invertebrates?
Should not harm any other invertebrates.

WILL fish pose a threat?
Fish may harass large or small individuals, pecking at their flesh or tentacles. Almost any species might have a go but angelfish are the usual suspects.

WILL it threaten fish?
No, at least not directly. If harassed, it may be goaded into releasing a purple fluid into the water. This is probably not toxic, but can reduce oxygen levels. In an already overstocked aquarium, such events can result in fish losses, mainly because fish have the largest demand for this most valuable resource.

SIMILAR SPECIES
The sea hare resembles a number of genera that feed on a similar diet. There are also a few specialised molluscs called saccoglossans that have an overall green coloration and consume algae. However, unlike the sea hare that uses the rasping 'teeth' on its radula to crunch algae, the saccoglossans pierce algal cells and suck out their contents.

WHAT to watch out for?
The sea hare is notorious for finding unprotected pump intakes and overflow holes in the aquarium, usually with disastrous results. If it is discovered quickly, the best strategy is to turn off the pump and if the slug has survived it can often crawl out of its own accord. In this case it has a good chance of survival.

WILL it reproduce in an aquarium?
Reproduction has been reported for this species and in some instances, where the aquarium is large enough, the resultant young have survived. Some species have been reported laying up to 85 million eggs in a single clutch!

Chelidoneura varians

Neon velvet sea slug

PROFILE

The neon velvet slug is not a nudibranch, but belongs to a group of molluscs called the Cephalaspidae, or head shield snails. It is commonly bought to control the nuisance flatworm *Convolutriloba retrogemma,* but is seldom successful in this role, as the flatworms reproduce faster than the slug can eat them.

WHAT size?
Large specimens reach 8cm or so.

WHAT does it eat?
Flatworms. This animal cannot be maintained without this food.

WHERE is it from?
Tropical Indo-Pacific.

WHAT does it cost?
★★☆☆☆   ★★★☆☆
Price depends on size and source.

SIMILAR SPECIES
There are several species within the genus *Chelidoneura,* all of which share the very specialised diet of *C. varians*.

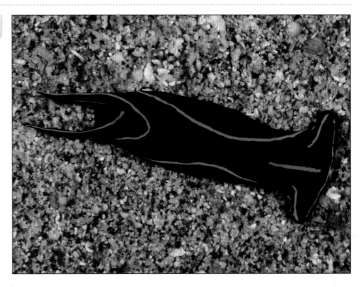

HOW many in one tank?
Keep singly to ensure that food does not become limited.

HOW compatible with other invertebrates?
No threat to any invertebrates, apart from its flatworm prey.

WILL fish pose a threat?
Very few fish will harass this slug. The colourful blue markings may advertise its distastefulness to predators.

WILL it threaten fish?
No. Completely harmless towards fish.

WHAT to watch out for?
This mollusc frequently falls victim to pump intakes and similar devices in the reef

▲ The distribution of the striking neon blue lines on a solid black body colour make this an unmistakable species of sea slug.

aquarium. Many sea slugs and their relatives are shortlived in captivity and their age at purchase is next to impossible to determine.

WILL it reproduce in an aquarium?
Yes. Sexual reproduction can and does occur where two individuals are housed. However, a single animal will deposit an egg mass (or a number of such masses) shortly before dying, as it slowly starves.

Elysia crispata

Lettuce sea slug

PROFILE

A stunningly beautiful animal that requires highly specialised care if it is to thrive. It is one of several species of sea slug known as saccoglossans. These slugs do not 'chew' their food; instead they use a stylet-like structure with a singe row of 'teeth' to puncture the cells of their algae food and suck out the contents. Their colour is the result of photosynthetic pigments and highly variable.

WHAT size?
Achieves around 5cm in length.

WHAT does it eat?
Consumes the contents of algal cells. It also contains photosynthetic chloroplasts that enable it to obtain nourishment from the illumination in the aquarium.

WHERE is it from?
Tropical Western Atlantic: Caribbean.

WHAT does it cost?
★★☆☆☆ ★★★☆☆
Medium priced.

▶ *The beautiful folded mantle that inspired the animal's common name is home to viable photosynthetic cells, garnered from the sea slug's diet of algal cell contents.*

HOW many in one tank?
Keep singly or in groups, where sufficient food allows.

HOW compatible with other invertebrates?
Should not harm any invertebrates. However, given its requirement for algae that are often not welcomed in the reef aquarium, housing this slug with corals may not be practical. Many aquarists maintain it in a well-illuminated sump aquarium that is home to macroalgae.

WILL fish pose a threat?
Some fish may nibble at the frilly folds on the dorsal surface. This slug is best kept in a fish-free aquarium, such as a refugium or sump aquarium containing algae.

WILL it threaten fish?
No.

WHAT to watch out for?
Specimens must be acclimatised over a prolonged period, ideally by drip-feeding aquarium water into a vessel containing the

SIMILAR SPECIES
The genus *Elysia* contains several species that are very similar in terms of appearance and husbandry. Some offered for sale may have specialised dietary requirements, which aquarists should research before making a purchase.

transportation water. This process can take several hours, during which time the water temperature must remain constant. As an aquarium animal, the lettuce sea slug does not generally do well in high-flow systems.

WILL it reproduce in an aquarium?
Healthy pairs will produce fertile eggs. Specimens are hermaphrodite. There are few accounts of larvae being raised successfully. Ailing specimens may also lay eggs, but these are usually infertile.

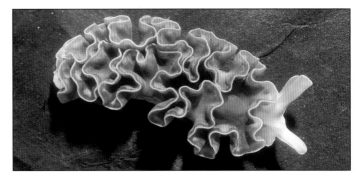

Lambis lambis

Spider conch

PROFILE

A large and interesting gastropod mollusc that is readily identified by the large, robust spines radiating from the edge of the shell. It is widely available, but in most aquariums it requires target-feeding if it is to thrive.

WHAT size?
Achieves around 15cm in length.

WHAT does it eat?
Eats large amounts of algae, but also consumes meaty particulate foods, such as brineshrimp and mysis. Place these items close to the animal to prevent its tankmates competing for them. Offer dried algae on a lettuce clip placed at the bottom of the aquarium to supplement the natural forms present in the aquarium. Algae tablets may also be accepted.

WHERE is it from?
Tropical Indo-Pacific.

WHAT does it cost?
★★★☆☆
Medium priced.

▲ *Large spider conchs are available in the hobby. Their unique shell shape is not well suited to confined spaces and they require regular feeding.*

HOW many in one tank?
Can be housed in numbers where space and food allow.

HOW compatible with other invertebrates?
May occasionally feed on invertebrates, particularly molluscs. This tends to be a problem if the spider conch is not adequately fed. It can knock over sessile invertebrates if they are not fixed in place.

WILL fish pose a threat?
This mollusc is safe with most reef-compatible fish.

◀ *Different species of* Lambis *can be identified from the number and shape of the radiating shell spines.*

WILL it threaten fish?
No.

WHAT to watch out for?
Specimens are often large, with appetites to match. They do best where they have plenty of open sandy spaces, although they will climb on rockwork.

WILL it reproduce in an aquarium?
This is possible, but the planktonic larvae are unlikely to survive in most aquariums.

SIMILAR SPECIES
Several species belong to the genus *Lambis* and most require very similar care. They also resemble the *Strombus* conchs.

Strombus luhuanus

Strawberry-lipped conch

PROFILE

A medium-sized mollusc, with several characteristics likely to endear it to the reef aquarist. Also known as the sand-shifting conch, it fulfils a useful role in the aeration of sand beds without consuming any of the beneficial flora and fauna that may occur there.

WHAT size?
Achieves a maximum size of about 9-10cm shell length.

WHAT does it eat?
This mollusc is particularly fond of diatoms, but also consumes various forms of algae commonly encountered in a marine aquarium. It is also said to scavenge uneaten fish food and detritus.

WHERE is it from?
Tropical Indo-Pacific.

WHAT does it cost?
★★☆☆☆
★★★☆☆
Moderately expensive for a mollusc, but well worth it.

HOW many in one tank?
Can be housed in numbers where there is available space and food.

HOW compatible with other invertebrates?
The snail's relatively large maximum size means that it has the potential to knock over specimen invertebrates. However, it is almost exclusively confined to the sand or gravel areas of the aquarium and will not usually climb rockwork unless it is very hungry.

WILL fish pose a threat?
Few reef-compatible fish bother the strawberry-lipped conch, especially as it often spends prolonged periods buried in sand.

WILL it threaten fish?
This species of marine snail has a modified operculum (the covering to the shell opening), which is sabre-like and serrated. Although

▼ *This hardy conch is highly useful for its burrowing and grazing habits.*

SIMILAR SPECIES
Several species of conch and closely related gastropod molluscs are offered for sale. One species that is perhaps more common than others is the fighting conch (*Strombus alatus*), which also proves hardy and long-lived in most aquariums.

the conch presents no threat to fish, it will thrash the operculum menacingly if it is attacked or threatened by them. Aquarists should also be aware of this defensive adaptation!

WHAT to watch out for?
The strawberry-lipped conch is a superb addition to most coral-rich aquariums in which sand is present, because they graze on films of algae growing on the surface of this medium. They will also plough through the sand or gravel, thereby aerating it. Both these jobs are helpful in maintaining a healthy aquarium without preventing the colonisation of the sand by beneficial bacteria and animals.

WILL it reproduce in an aquarium?
Courtship and mating may be observed in an aquarium situation. Several species of *Strombus* conch have been aquacultured and this is a reasonable candidate for a home breeding project.

Cypraea tigris

Tiger cowrie

PROFILE

The tiger cowrie is one of the most unmistakable species of mollusc available in the hobby. The cryptic markings on its shell actually resemble those of the leopard, rather than the big cat that lends the animal its common name. The beautiful shell is obscured when the animal extends its mantle over the surface. This thin sensory structure with numerous fleshy extensions gives the mollusc some ability to detect predators before they can launch an attack.

WHAT size?
Grows to around 10cm shell length.

WHAT does it eat?
This omnivorous mollusc increases its dietary range as it ages. Large specimens will consume anything from algae to dead fish. Offer the animal shellfish throughout its time in the aquarium. Dried algae also helps to provide adequate nutrition.

WHERE is it from?
Tropical Indo-Pacific.

WHAT does it cost?
★★☆☆☆ ★★★☆☆
Price will depend on whether the mollusc has been imported directly or through a wholesaler.

▲ *The tiger cowrie is one of the most unmistakable species of mollusc available in the hobby.*

HOW many in one tank?
Best kept singly unless sufficient food can be guaranteed.

HOW compatible with other invertebrates?
Larger specimens will consume anemones, soft corals, colonial polyps and other encrusting animals. Increasing the amount of food available to the snail might delay this behaviour, but it is only a temporary reprieve.

WILL fish pose a threat?
Some curious fish may occasionally peck at the mantle of this cowrie when it is extended. Dwarf and true angelfish, butterflyfish and wrasse sometimes find this too tempting to resist. The snail is able to move with the mantle retracted, which may be a clue to some hitherto unnoticed harassment by a fish.

WILL it threaten fish?
This snail presents no threat to any fish species. In fact, it is a reasonable choice as scavenger/herbivore for a live rock-based fish-only aquarium housing fairly benign fish, providing the system does not experience high nitrate levels.

WHAT to watch out for?
Before buying them, give specimens time to settle after importation. Choose individuals that are moving about with their mantles expanded.

WILL it reproduce in an aquarium?
Although sexual reproduction is extremely rare, it cannot be ruled out. However, the cowrie is not maintained in many reef aquariums and, where it is present, only single specimens are represented.

SIMILAR SPECIES
Several species of cowrie make their way into the hobby, some more regularly than others. Their diet is likely to be similar to that of the tiger cowrie, but some species are entirely herbivorous and represent no threat to a reef aquarium. Always establish the diet of potential purchases before buying them.

Hippopus hippopus

Bear's paw clam

Although not as stunning as members of the genus *Tridacna*, the bear's paw clam can prove very hardy and long lived in the reef aquarium. Its mantle does not overlap the shell. It is most often encountered on natural reefs, occupying sandy substrates in relatively shallow water.

WHAT size?
Achieves around 40cm shell length on average, although larger individuals have been reported.

WHAT does it eat?
The mantle contains photosynthetic algae that provide the animal with a significant amount of nourishment. However, in common with all bivalve molluscs, it has enlarged gills that capture food as well as enabling the animal to breathe. Offering phytoplankton and microplankton, such as oyster eggs, will supplement the diet.

WHERE is it from?
Tropical Indo-Pacific.

WHAT does it cost?
★★★☆☆ ★★★★☆
Price depends on size and origin.

HOW many in one tank?
Can be housed in numbers provided sufficient space for growth and suitable substrates are on offer.

HOW compatible with other invertebrates?
Not harmful to any aquarium invertebrates.

WILL fish pose a threat?
The lack of an overhanging mantle means that this species is less likely to be nipped by fish than other aquarium clams.

WILL it threaten fish?
No.

WHAT to watch out for?
Ensure that individuals have an undamaged mantle. Avoid specimens with clear patches of mantle.

WILL it reproduce in an aquarium?
Spawning is possible. This

▲ Hippopus *clams may not share the stunning good looks of clams from the genus* Tridacna, *but they are attractive and interesting in their own right.*

species releases eggs and sperm into open water. Although the clams are hermaphrodite, they do not self-fertilise, so synchronous spawning of two individuals must occur for fertilisation to be possible. In any case, it is unlikely that the planktonic larvae will survive.

SIMILAR SPECIES
The closely related *H. porcellanus* is very similar in appearance, but lacks the thin red lines on each valve present in *H. hippopus*. Encrusting organisms may obscure these markings, but husbandry for both species is practically identical.

Tridacna derasa

Derasa, or smooth-shelled, clam

PROFILE

This beautiful species of clam is probably the hardiest of the members of the genus *Tridacna* commonly available in the hobby. It has fewer colour varieties than some, but is a spectacular addition to the reef aquarium. It is found in turbid areas of reefs, usually on sandy substrates, and consequently has a less well developed byssus gland than its close relatives.

WHAT size?
Can achieve a shell length of 30cm or more.

WHAT does it eat?
The mantle contains photosynthetic algae that will provide the animal with a significant amount of nourishment. However, in common with all bivalve molluscs, it has enlarged gills that capture food as well as enabling the animal to breathe. Offering phytoplankton and microplankton, such as oyster eggs, will supplement the diet.

WHERE is it from?
Tropical Indo-Pacific.

WHAT does it cost?
★★☆☆☆   ★★★★☆
Price depends on size and origin.

HOW many in one tank?
Allow for the growth of this clam. It is possible to house more than one, but they will need plenty of space.

HOW compatible with other invertebrates?
Should not harm other invertebrates. The mantle is sensitive to contact with any stinging animals and needs plenty of room to grow into.

WILL fish pose a threat?
Some fish will nip at the mantle of clams. True and dwarf angelfish are particularly guilty of such behaviour, but it is difficult to identify particular species as it may depend on the individual concerned.

WILL it threaten fish?
No.

WHAT to watch out for?
Select specimens with mantles that extend over the edge of the shell. Ignore individuals that

THE BYSSUS GLAND
This is the gland that secretes the anchoring threads that hold the clam in place. Many aquarists place specimens on sand or gravel to make moving them easier, as they will bind tightly to rocks.

SIMILAR SPECIES
The mantle is similar to that of *Tridacna squamosa*, but the latter has obvious extensions (scutes) on the margins of each valve. *T. maxima* and *T. crocea* also have rough shells, with or without scutes, but generally more colourful mantles.

appear to have pale patches punctuating the colourful pigments.

WILL it reproduce in an aquarium?
Although not a common experience, clams can spawn in captivity. They have sometimes been induced to spawn with hormonal injections. Has been successfully reared in captivity.

◀ *Derasa clams are less variable in their mantle colour and patterning than species such as* T. maxima *and* T. crocea*. An extended mantle as seen here is a sign of good health.*

Tridacna maxima

Maxima clam

If it is to thrive, this bivalve mollusc must have excellent water quality and strong lighting. Clams from the genus *Tridacna* have symbiotic algae in their mantle tissue, which supply them with nourishment. They are also able to filter small particulate material from the water.

WHAT size?
Large specimens reach 30-35cm or so.

WHAT does it eat?
Primarily photosynthetic, but the clam also welcomes suspension feeds intended for filter-feeding invertebrates. It is a good idea to add phytoplankton for small specimens measuring 8cm or less, but clams of all sizes will benefit from its introduction.

WHERE is it from?
Tropical Indo-Pacific.

WHAT does it cost?
★★★☆☆　★★★★★
Price depends on size, source and colour. Cultured specimens are widely available; seek these out wherever possible.

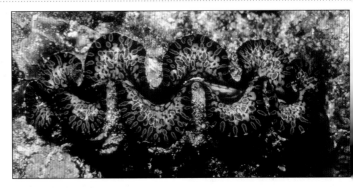

HOW many in one tank?
Can be kept in numbers but allow room for growth.

HOW compatible with other invertebrates?
Maxima clams do not pose any threats to other invertebrates, but they themselves are vulnerable to stinging neighbours, predatory worms and snails. The latter usually arrive by accident on coral base rock or live rock.

WILL fish pose a threat?
Notorious 'mantle nippers' include butterflyfish and many species of dwarf angelfish *(genus Centropyge)*. Some true angelfish are also untrustworthy. Fish that move quantities of sand with their mouths may irritate clams with their excavations.

WILL it threaten fish?
No. This species presents no dangers towards any fish.

WHAT to watch out for?
Check for damage to the byssus gland (see page 145).

▲ *The mantle of this specimen clearly shows a sample of the colours that are encountered in* T. maxima.

WILL it reproduce in an aquarium?
Maxima clams have been reported spawning in the aquarium, but it is virtually impossible for home aquarists to care for the planktonic larvae. Has been successfully reared in captivity in controlled conditions.

SIMILAR SPECIES
The boring clam *(Tridacna crocea)* and squamose clam *(T. squamosa)* are similar in appearance and husbandry. The true giant clam *(T. gigas)* is sporadically available and similar in terms of its aquarium requirements. All species enjoy strong illumination.

Lima spp.

Flame scallop

PROFILE

This bivalve mollusc is regularly imported for the aquarium trade and cheap to buy, but requires a lot of attention if it is to survive for any length of time. Aquarists are often tempted by its vivid coloration and the 'electric' flashes of the hairlike cilia that beat rhythmically to induce flow over the enlarged, plankton-consuming gills. Quite why the scallop draws attention to itself with this adaptation is unknown.

WHAT size?
May reach 10cm shell length.

WHAT does it eat?
This scallop feeds exclusively on planktonic organisms and must have similar substitutes on a daily basis to satisfy its needs in the aquarium. Phytoplankton is useful, together with offerings such as oyster eggs, 'marine snow' and the widely available small zooplankton substitutes.

WHERE is it from?
Lima species are found in the Tropical Indo-Pacific and Caribbean.

WHAT does it cost?
★★☆☆☆
Price depends on size and origin.

HOW many in one tank?
Can be kept singly or in numbers, provided sufficient food can be guaranteed.

HOW compatible with other invertebrates?
Does not represent a threat to other invertebrates. Some starfish attempt to eat scallops, but such species are generally not included in the livestock of a reef aquarium. The scallop is able to evade capture by rapidly opening and closing the valves of its shell, creating water jets that propel it backwards away from the potential danger.

WILL fish pose a threat?
Occasionally, fish show more than a passing interest in this animal and nibble at its tentacles. Dwarf angelfish are particularly guilty of this behaviour. Some *Lima* species have sticky tentacles that detach and adhere to their predator, much to the latter's annoyance.

WILL it threaten fish?
No.

WHAT to watch out for?
Although watching a scallop clapping its shell and 'swimming' through the water is fascinating to observe, it rapidly exhausts the animal's resources. Never goad a scallop into this response in the aquarium; it

▲ *Flame scallops often locate in areas where they can attach to a firm substrate with favourable currents.*

can take weeks to recover fully. Scallops are capable of moving around in the aquarium and attaching to suitable hard substrates, often where there are good currents. Choose specimens with expanded tentacles and strong coloration.

WILL it reproduce in an aquarium?
Reproduction cannot be ruled out, especially if specimens are receiving enough food.

SIMILAR SPECIES
At least three or four species of scallop that open their valves wide and extend long tentacles make their way into the hobby. The aquarium requirements for all are very similar. The key to success lies in providing them with regular supplies of suitable foodstuffs.

Perna viridis

Green-lipped mussels

PROFILE

Despite being hardy and invasive in tropical seas, green-lipped mussels are not easy to maintain in a reef aquarium. The secret of success lies in supplying the correct food in sufficient quantities. The mussels make an attractive addition to the larger aquarium.

WHAT size?

Each shell will grow to about 12cm long, but the colony will not increase in number.

WHAT does it eat?

Accepts phytoplankton and very fine zooplankton, such as oyster eggs. The animal becomes more selective with larger particles, but the availability of off-the-shelf diets containing rotifers and similar zooplankton, plus 'marine snow', means that aquarists have every chance of success with this bivalve.

WHERE is it from?

Native to African and South American waters, its has invaded Southeast Asia and the tropical Atlantic.

WHAT does it cost?

★★★☆☆

The price is per colony.

▶ _Bivalve mollusc shells often harbour encrusting animals such as tubeworms or, as here, barnacles._

HOW many in one tank?

Usually bought in a colony of four or more individuals.

HOW compatible with other invertebrates?

Specimens may leave the main colony and attach to other substrates using their byssus threads. This may cause irritation to other invertebrates.

WILL fish pose a threat?

Most reef-compatible fish ignore this species. Avoid butterflyfish such as the copperband _(Chelmon rostratus)_. Dwarf

SIMILAR SPECIES

There are at least three other species of green mussel. Some may be sub-tropical in origin and may not thrive, even when given sufficient food.

angelfish may nip at the small amount of exposed mantle present when the shell is open.

WILL it threaten fish?

No.

WHAT to watch out for?

Choose specimens that are well settled in the dealer's tank. All the individual shells should be intact and closed. Acclimatise specimens over a prolonged period. Ailing specimens have the potential to pollute the aquarium.

WILL it reproduce in an aquarium?

Spawning may occur when the colony is provided with sufficient suitable foods, but the larvae are planktonic and unlikely to survive for long in most reef aquariums.

Spondylus princiceps

Pacific thorny oyster

PROFILE

Although it is possible to buy specimens, usually through suppliers obtaining livestock from the Philippines, this bivalve mollusc is often acquired as the substrate to a colonial polyp or soft coral. The encrusting life forms that develop on the valves of the mollusc make it look like a lump of rock until it opens its shell, revealing a spectacular mantle.

WHAT size?
Grows to about 15cm long.

WHAT does it eat?
The large gills sort very fine particulate material from inedible particles. For the best results, offer phytoplankton, oyster eggs and a variety of 'off-the-shelf' plankton substitutes on a daily basis.

WHERE is it from?
Tropical Pacific.

WHAT does it cost?
★★★☆☆

SIMILAR SPECIES

The Atlantic thorny oyster *(Spondylus americanus)* is occasionally offered for sale. Other *Spondylus* species are sometimes available; all require similar care. Some of these have spectacular mantle coloration.

HOW many in one tank?
Can be kept in groups, but usually encountered singly in its natural environment.

HOW compatible with other invertebrates?
Offers no threat to any invertebrates in most circumstances. However, if the animal dies as a result of poor maintenance, it has the potential to pollute an aquarium.

WILL fish pose a threat?
Species with a reputation for nipping clam mantles, such as dwarf angelfish *(genus Centropyge),* may also have a nibble at the less exposed flesh of this bivalve. However, the oyster can quickly close its shell when danger threatens.

WILL it threaten fish?
No.

▲ *The large paired gills typical of the filter-feeding bivalve molluscs can be clearly seen through the opening in the mantle of this thorny oyster.*

WHAT to watch out for?
The encrusting life found on the surface of this animal may not survive in an aquarium. Sponges are particularly sensitive; if they die and begin to decompose they can pollute the oyster's local environment and stress or potentially kill it and any neighbours. Remove them carefully before introducing the oyster into the tank.

WILL it reproduce in an aquarium?
Spawning may occur, but the larvae are planktonic and unlikely to survive for long in most reef aquariums.

Fancy filter feeders

Marine aquarists establishing aquariums along current preferred lines, with live rock as the substrate for biological filtration, will inevitably experience polychaete worms, which frequently arrive in this medium by accident. This reflects the fact that the 12,000 species of mostly marine worms (Phylum Annelida) have been recorded living in huge densities in tropical reefs. Paradoxically, aquarists are offered only a small number of species. These are the fanworms, which have specialised feeding crowns around their mouths, evolved to capture particulate material from the water column. Historically, many species were difficult to keep long term as aquarists struggled to feed them. However, developments in invertebrate foods, namely zooplankton and phytoplankton substitutes and additives, have greatly increased the chances of success with these animals.

Price guide

★	£5 – 15
★★	£15 – 20
★★★	£20 – 30
★★★★	£30 – 40
★★★★★	£40 – 50

Although it may contain a number of worm individuals, each 'cluster', or colony, usually measures no more than 10cm in diameter.

Bispira brunnea

Featherduster cluster

HOW many in one tank?
Can be kept in groups of any size, providing you can supply them with sufficient food.

HOW compatible with other invertebrates?
Will not harm other invertebrates.

WILL fish pose a threat?
Avoid butterflyfish, including copperbands *(Chelmon rostratus)*. Dwarf angelfish *(Centropyge* spp.*)* may also nip at the crowns of this species.

WILL it threaten fish?
No, harmless towards fish.

WHAT to watch out for?
The key to success with this species is ensuring that it is not harassed by other tankmates and that it obtains sufficient food. It will also benefit from a stable position in the aquarium, such as in the substrate (with crowns protruding) or inside a crack or crevice.

WILL it reproduce in an aquarium?

SIMILAR SPECIES

Not all members of the genus *Bispira* are colonial, but their identification can be difficult. Some will thrive in a marine aquarium, especially when 'planted' directly into the substrate.

Asexual reproduction is common, provided the colony is obtaining sufficient food. Sexual reproduction is possible given adequate nourishment.

Filograna implexa

Coral worm

A little-known species of colonial tubeworm with beautiful red-and-white feeding crowns that makes an interesting inclusion in a reef aquarium. The long tubes made of calcium carbonate are thin yet surprisingly robust, providing they are carefully handled.

WHAT size?
The worm colonies can measure perhaps 15-20cm in length, but each feeding crown is very small, being less than 1cm across.

WHAT does it eat?
This tubeworm requires fine particulate feeds, such as microplankton or similar substitutes. Phytoplankton is also likely to be a useful regular addition.

WHERE is it from?
Circumtropical. Most imported specimens have been collected from the Philippines and Indonesia.

WHAT does it cost?
★★☆☆☆   ★★★☆☆
Price depends on the size of the colony and its place of origin.

▲ *Before buying this attractive species observe the colony closely, as many of the tubes are likely to be unoccupied.*

HOW many in one tank?
Specimens are acquired as colonies of varying numbers of worms.

HOW compatible with other invertebrates?
Will not harm other invertebrates.

WILL fish pose a threat?
Many fish will find the prospect of removing the feeding crown of a tubeworm too tempting to resist. Butterflyfish and dwarf angelfish can be particularly interested in these worms.

WILL it threaten fish?
No. Harmless towards fish.

WHAT to watch out for?
Provide plenty of water movement. Take care when selecting a specimen. Try to ensure that as many feeding crowns as possible are visible in the colony. Check for any signs of algal growth, which can discourage the worm from emerging to feed.

WILL it reproduce in an aquarium?
Reports of sexual reproduction are scarce, largely because this worm is less familiar to aquarists than some other polychaete tubeworms. This is a reflection of the numbers currently maintained in reef aquariums.

SIMILAR SPECIES
There are several species of *Filograna* present in almost all saltwater environments around the globe. Due to the distinct form of the tubes, this group is unlikely to be confused with other so-called 'hard tubeworms', such as the koko worm (*Protula bispiralis*).

Protula bispiralis

Koko worm

FISH PROFILE

This is a species of polychaete fanworm known as a 'hard tubeworm' because it secretes its own tube made from calcium carbonate. Specimens can have different-coloured feeding fans; white, pink and red-orange are the most commonly found.

WHAT size?
The hard calcium carbonate tube can reach 60cm or more, with the paired crowns measuring around 3.5-4cm each.

WHAT does it eat?
Bacteria, phytoplankton and zooplankton all form part of the diet of this sessile worm. Similar offerings should be made in the aquarium, target feeding where possible. 'marine snow'-type products are also likely to be useful.

WHERE is it from?
Tropical Indo-Pacific.

WHAT does it cost?
★★☆☆☆   ★★★★★
Cost usually depends on the colour morph, with deep-red specimens being the most desirable and expensive. White-crowned worms are generally the cheapest.

▶ *Red- or pink-crowned specimens of koko worm are justifiably pricey.*

HOW many in one tank?
Can be kept in groups of any size, providing you can supply them with sufficient food.

HOW compatible with other invertebrates?
Will not harm other invertebrates.

WILL fish pose a threat?
Dwarf angelfish, butterflyfish and some wrasse will nip at the tentacle crown, causing it to remain retracted. It is therefore unable to feed and will starve if you do not take action to remove the nuisance fish.

WILL it threaten fish?
No, harmless towards fish.

WHAT to watch out for?
Specimens often shed the tentacle crown shortly after being imported or relocated to a new aquarium. This can also

SIMILAR SPECIES
Worms that secrete a rigid tube are unable to move around the aquarium in the same way as their more flexible relatives (such as *Sabellastarte* spp.). It is very important that they are positioned in a suitable region of the aquarium where they can be fed easily and are subjected to moderate- to high-flow rates.

occur during times of stress, perhaps due to elevated water temperature or starvation. Rarely, the entire animal may exit the tube, in which case its chances of survival are poor.

WILL it reproduce in an aquarium?
This species is reported to spawn frequently. It releases sperm and eggs directly into the water column. The resulting larvae are unlikely to survive.

Sabellastarte spp.

Featherduster worm

FISH PROFILE

Perhaps the most commonly available tubeworm and a firm favourite with aquarists because of its good looks and low price. However, it is not necessarily easy to maintain, unless provided with the correct diet.

WHAT size?
The tube may be up to 15cm long. The feeding apparatus, or head, may be 10cm or so in diameter.

WHAT does it eat?
The 'featherduster' part of the worm is actually a sophisticated plankton-capturing device. There are several good-quality phytoplankton- and zooplankton-based foods available in the aquarium hobby and these should be added regularly to the aquarium, otherwise the worm will starve.

WHERE is it from?
Tropical seas around the world.

WHAT does it cost?
★☆☆☆☆
Inexpensive.

▶ *There are several species of worm known as featherdusters. Although the husbandry for each is similar, some can prove more hardy in an aquarium than others.*

HOW many in one tank?
Can be kept in groups of any size, providing you can supply them with sufficient food.

HOW compatible with other invertebrates?
Should not harm other invertebrates.

WILL fish pose a threat?
Butterflyfish, including the copperband *(Chelmon rostratus)*, find most species of tubeworm too tempting not to nip at. Although this behaviour is unlikely to kill the worm, it can cause it to remain retracted. As a result, the worm cannot feed and will usually starve over a period of months.

WILL it threaten fish?
No, harmless towards fish.

WHAT to watch out for?
Repeated shedding and regrowing of the heads can be a sign of starvation. Each head can be a little smaller than the previous one. Tubes are constructed from particles of sand and gravel in the aquarium substrate and are constantly being eroded and replaced. This makes it very difficult to move them without exposing the worm.

WILL it reproduce in an aquarium?
Potentially. This is a broadcast-spawning species that releases sperm and eggs directly into the water column.

Spirobranchus spp.

Plume rock, or Christmas tree worm

FISH PROFILE

This species is unusual in that it is commonly found in association with the stony coral *Porites* sp. The worms settle as larvae on the coral and their tubes become overgrown by it. It is widely believed that the two animals have some sort of symbiotic relationship and it is certainly true that the coral does not thrive without the worms and vice versa.

▲ *It is vital to determine the health of both the worms and the coral before buying. Check that the worms are at home in their tubes and the coral's polyps are expanded.*

WHAT size?
Each worm can reach 7cm in length, but this is obscured by the coral. Feeding crowns achieve 3cm at most, but the coral itself has almost infinite growth potential.

WHAT does it eat?
Many colonies apparently thrive without deliberate feeding attempts by the aquarist, suggesting that the coral host may influence their diet. However, they are likely to benefit from regular, target-fed additions of phytoplankton and small zooplankton. The coral requires strong flow and illumination.

WHERE is it from?
Tropical Indo-Pacific.

WHAT does it cost?
★★★☆☆ ★★★★★
Price depends on the size of the host coral.

HOW many in one tank?
Can be kept in groups of any size, providing you can supply them with sufficient food.

HOW compatible with other invertebrates?
Will not harm other invertebrates.

WILL fish pose a threat?
The brightly coloured crowns of this species prove too tempting to many fish, including butterflyfish and some species of dwarf angelfish. Small, otherwise reef-safe wrasse might also attack them, causing the worms to remain retracted, sometimes only emerging at night.

TITLE
Spirobranchus tubeworms are sometimes found in live rock or in association with other species of hard coral.

WILL it threaten fish?
No, harmless towards fish.

WHAT to watch out for?
Generally speaking, the key to success with plume rock is providing the coral with suitable conditions in terms of lighting and flow. A healthy coral will often mean healthy tubeworms.

WILL it reproduce in an aquarium?
This species does not apparently reproduce asexually, so numbers of worms will not increase in a given piece of coral. It may spawn sexually if it receives adequate nutrition, but the larvae are planktonic and will not survive in the aquarium.

Symmetrical stunners

▷ Echinoderm translates as 'spiny skin', a feature shared by most of these animals. However, a characteristic they all have in common is their particular form of radial symmetry called pentamerous symmetry. This means that the body can be divided into five similar parts around a central axis. We know these animals as starfish, sea urchins and sea cucumbers. All the 6,000 described species of echinoderm (Phylum Echinodermata) have a water-vascular system that powers their tube feet. These specialised structures are also

Price guide

★	£4 – 10
★★	£10 – 20
★★★	£20 – 30
★★★★	£30 – 40
★★★★★	£40 – 45

used in gas exchange, feeding and a host of other tasks. The system appears to be sensitive to sudden changes in water chemistry, particularly salinity. To prevent undue stress or damage to these animals, aquarists are strongly advised to acclimatise any echinoderm over a prolonged period before introducing it into the aquarium.

PROFILE

Although fairly uninspiring to look at, with the correct management this starfish can become a useful addition to some aquariums. It lives primarily in sand or rubble habitats, where it spends most of its time buried. In a reef aquarium its constant movement through the substrate can keep sand beds clean and aerated. However, it will consume beneficial animals that have colonised this area.

WHAT size?
Grows to about 20cm across the arms.

WHAT does it eat?
Consumes detritus that accumulates in sand or gravel. Will eat the living contents of the sand substrate before becoming more adventurous and taking on pieces of shellfish, mysis and brineshrimp. It may be necessary to place food directly beneath the animal to encourage it to feed.

WHERE is it from?
Tropical Indo-Pacific.

WHAT does it cost?
★☆☆☆☆   ★★☆☆☆
Inexpensive. The upper limit of the price range reflects the largest specimens, but most are much less expensive.

Asteropecten spp.

Sand-shifting starfish

HOW many in one tank?
It is usual to keep more than one specimen in an aquarium, but the animals will rapidly consume all the natural life present. At this point, each starfish will require individual feeding if it is to survive.

HOW compatible with other invertebrates?
Apart from its obvious impact on the animals that naturally colonise the sand or gravel in a live rock-based aquarium, this starfish does not harm any other invertebrates.

SIMILAR SPECIES

There are several species of starfish belonging to the genus *Asteropecten* that live in sandy substrates. They are characterised by rows of toothlike projections along the sides of each arm.

WILL fish pose a threat?
Few fish will bother this starfish. Angelfish and butterflyfish might attempt to nip at its tube feet.

WILL it threaten fish?
No, it offers no threats to any fish in the aquarium, although it may scavenge dead specimens.

WHAT to watch out for?
Only select animals that have been settled into the dealer's aquarium for a week or more. These should have recovered from the rigours of shipping.

However, specimens that have been kept for prolonged periods may be near starvation, in which case they often begin to disintegrate. This situation is reversible, providing the starfish is offered suitable foods.

WILL it reproduce in an aquarium?
No. It may spawn, but this is unlikely to be successful.

▼ *The movement of the sand-shifting starfish keeps the sand turned over.*

SEPARATE SEXES

Echinoderms are either male or female. Thus it takes two to tango in the home aquarium and they must be of different genders. Therefore, asexual reproduction is more commonly experienced.

Echinaster luzonicus

Orange starfish

This beautiful species is regularly imported via Sri Lankan suppliers. It resembles species from the genus *Linckia,* and its true affinities to any one group are the cause of some debate, although it is most commonly assigned to the genus *Echinaster.* Close scrutiny of the orange-red tissue reveals small black spots that are useful in separating this starfish from similar species.

WHAT size?
Grows to about 15cm across the arms.

WHAT does it eat?
Its diet appears to include algae, sponges and encrusting organisms found on live rock. To satisfy its needs, stock it into a mature aquarium with plenty of live rock. Encourage the starfish to feed by carefully placing it on top of chopped shellfish or mysis and algae.

WHERE is it from?
Tropical Indian Ocean, but may be found throughout the Pacific.

WHAT does it cost?
★★☆☆☆   ★★★☆☆
Moderately expensive, depending on overall size and whether it has been directly imported or acquired via a wholesaler.

HOW many in one tank?
Best kept singly to increase the availability of naturally occurring foods in the aquarium.

HOW compatible with other invertebrates?
This starfish may occasionally target bivalve molluscs such as clams, but this is rare.

WILL fish pose a threat?
Few fish will bother this starfish. Angelfish and butterflyfish might attempt to nip at its tube feet.

WILL it threaten fish?
No, it offers no threats to any fish in the aquarium, although it may scavenge dead specimens.

WHAT to watch out for?
Although less sensitive than many starfish to the rigours of shipping and subsequent acclimatisation to the new aquarium, it is a good idea to allow the orange starfish plenty of time in the dealer's tank before buying it. However,

bear in mind that often it will not be provided with sufficient food during this time. Until it is sold it is best kept with corals, so that it can browse the natural life on the base rock.

WILL it reproduce in an aquarium?
No. It may spawn but this is unlikely to be successful.

◀ *This is a highly variable species and individuals with multiple arms are not uncommon.*

Fromia indica
Red starfish

PROFILE

The red starfish is frequently imported through Sri Lankan suppliers and familiar to most hobbyists. However, it is not straightforward to maintain, as it is very sensitive to changes in water chemistry and often succumbs rapidly to bacterial infections resulting from rough handling, poor acclimatisation and starvation. Only stock it into a stable aquarium that has been established for at least 12 months.

WHAT size?
Grows to about 8-10cm across the arms.

WHAT does it eat?
The diet of this starfish is known to include algae and microorganisms that occur naturally on live rock and coral base rock. Offer a variety of foods and encourage the starfish to feed by placing it directly onto individual items.

WHERE is it from?
Tropical Indo-Pacific. Imported through Sri Lanka in large numbers.

WHAT does it cost?
★★☆☆☆
Relatively inexpensive, although larger specimens can demand a higher price.

HOW many in one tank?
Best kept singly to increase the availability of naturally occurring foods in the aquarium.

HOW compatible with other invertebrates?
Unlikely to bother any aquarium invertebrates. May irritate corals by crawling over them, but seldom does any damage to the victim of this clumsiness.

WILL fish pose a threat?
Few fish will bother this starfish. Angelfish and butterflyfish might attempt to nip at its tube feet.

WILL it threaten fish?
No, it offers no threat to any fish in the aquarium.

WHAT to watch out for?
Never buy recently imported specimens, as they need time to settle and acclimatise properly following the rigours of shipping. Avoid specimens showing any signs of damage, often indicated by powdery

▲ *Fromia indica is variable in form and colour, with red specimens being the most highly prized.*

deposits around the animal, or obvious open wounds. The arms should end in rounded tips.

WILL it reproduce in an aquarium?
Unlikely.

SIMILAR SPECIES

This is a variable species in which the red colour can appear orange or even brown or combinations of all of these colours. Several members of the genus *Fromia* offered for sale share a very similar body plan, differing only in coloration and patterning, as well as in the presence or absence of spots or scales on the dorsal surface.

Fromia monilis

Maldive, or elegant, starfish

A beautiful starfish, imported regularly through Sri Lanka. This may explain one of its common names, the Maldives being the major embarkation point for marine livestock from the surrounding area. The body is flattened, as in the biscuit starfish *(Pentagonaster duebeni)*, but it is much more flexible and often found 'hugging' the substrate as it searches for food.

WHAT size?
Grows to about 10-12cm across the arms.

WHAT does it eat?
In the wild, this starfish is believed to feed on sponges. Most of the suitable food in an aquarium is likely to develop in a mature system containing an abundance of live rock that is over 12 months old. Offering proprietary frozen foods intended for angelfish and containing sponge material is certainly worth a try, but the starfish must be placed directly on top of such offerings if it is to feed.

WHERE is it from?
Tropical Indo-Pacific.

WHAT does it cost?
★★☆☆☆
Price depends on the size of the individual concerned.

▲ *Handle F. monilis with care. In common with many tropical starfish, it is prone to bacterial infections and damage to its water vascular system.*

HOW many in one tank?
Best kept singly to increase the availability of naturally occurring foods in the aquarium.

HOW compatible with other invertebrates?
Unlikely to bother any aquarium invertebrates. May irritate corals by crawling over them, but seldom does any lasting damage.

WILL fish pose a threat?
Few fish will bother this starfish. Angelfish and butterflyfish might attempt to nip at its tube feet.

WILL it threaten fish?
No. It offers no threat to any fish in the aquarium.

WHAT to watch out for?
Never buy recently imported specimens, as they need time to settle and acclimatise following the rigours of shipping. Avoid specimens showing any signs of damage, often indicated by powdery deposits around the animal, or obvious open wounds.

WILL it reproduce in an aquarium?
Unlikely.

SIMILAR SPECIES
The starfish *Fromia nodosa* is similar in that it also has a mosaic of scales of a different colour to the rest of the animal around its centre. However, it also has nodules that run the length of the arms to their tips.

Linckia laevigata

Blue starfish

PROFILE

Being regularly imported and widely available, this is the most recognisable of the tropical marine starfish offered for sale. There is a red morph that is encountered infrequently, but its aquarium demands are identical.

WHAT size?
Grows to about 15cm across the arms.

WHAT does it eat?
Its diet appears to include algae, sponges and encrusting organisms found on live rock, as well as bacteria films. It will often target newly stocked aquarium corals, crawling all over their base rock in search of food. Stock it into a mature aquarium with plenty of live rock in order to supplement offerings such as dried algae and chopped shellfish. Placing these items directly beneath the starfish enables it to evert its stomach onto them.

WHERE is it from?
Tropical Indo-Pacific. Imported from the Philippines and Indonesia on a regular basis.

WHAT does it cost?
★★☆☆☆
Relatively inexpensive, although larger specimens can demand a higher price.

HOW many in one tank?
Best kept singly to increase the availability of naturally occurring foods in the aquarium.

HOW compatible with other invertebrates?
Unlikely to bother any aquarium invertebrates directly, but may dislodge poorly anchored specimens.

WILL fish pose a threat?
Few fish will bother this starfish. Angelfish and butterflyfish might attempt to nip at its tube feet.

WILL it threaten fish?
No, it offers no threat to any fish in the aquarium.

▼ *Linckia starfish are aquarium favourites. Blue is by far the most common colour variety, but orange or red ones can be found occasionally.*

SIMILAR SPECIES
The red form could be confused with starfish from the genera *Echinaster* or *Leiaster*, but the blue morph is unlikely to be mistaken for any other starfish.

WHAT to watch out for?
Specimens should be well settled in the dealer's tank before you buy them. They do not respond well to sudden changes in water chemistry, so careful acclimatisation over a prolonged period is essential. Watch out for the parasitic snail *Thyca crystalline* that feeds on the flesh of this animal. It resembles one half of the shell of a bivalve mollusc and in most cases measures less than 5mm. It is transparent, so it appears blue when seen on this starfish.

WILL it reproduce in an aquarium?
It can drop an arm, which has the potential to form a new animal given plenty of food. It may spawn, but this is unlikely to be successful.

Linckia multiflora

Multicoloured starfish

Despite displaying a variety of beautiful colours, aquarists often overlook this species. This is because the combination of pigments on its dorsal surface serve to break up its outline, so that in most tanks it simply merges into the background. However, by observing it closely, you can still enjoy its beauty.

WHAT size?
Grows to about 15cm across the arms.

WHAT does it eat?
Its diet appears to include algae, sponges and encrusting organisms found on live rock. Stock it into a mature aquarium with plenty of live rock to satisfy its needs. Offer chopped shellfish or mysis and algae, carefully placing the animal on top of one of these items to encourage it to feed.

WHERE is it from?
Tropical Indo-Pacific. Many specimens imported for the aquarium trade arrive from Sri Lankan exporters.

WHAT does it cost?
★★☆☆☆   ★★★☆☆
Relatively inexpensive at small sizes, although larger specimens can command a higher price.

◀ *This colourful starfish often has arms of differing lengths. This is an entirely natural state and not the result of poor handling, infection or any physical damage.*

HOW many in one tank?
Best kept singly to increase the availability of naturally occurring foods in the aquarium.

HOW compatible with other invertebrates?
Unlikely to bother any aquarium invertebrates.

WILL fish pose a threat?
Few fish will bother this starfish. Angelfish and butterflyfish might attempt to nip at its tube feet.

SIMILAR SPECIES
Unlikely to be confused with any other starfish, although some species from the genera *Nardoa* and *Echinaster* have a similar multicoloured appearance.

WILL it threaten fish?
No, it offers no threat to any fish in the aquarium.

WHAT to watch out for?
Choose well-settled specimens that have been in the dealer's tank for over a week to ensure that they have been well acclimatised and are fully recovered from the rigours of shipping. Dripping water from the aquarium into their transportation bag is a good idea, as they are sensitive to sudden changes in water chemistry. Avoid specimens with signs of discoloration on the body or any open wounds, particularly at the extreme margins of the arms.

WILL it reproduce in an aquarium?
It can drop an arm, which has the potential to form a new animal given plentiful food. It may spawn but this is unlikely to be successful.

Pentagonaster duebeni

Red blotch biscuit starfish

PROFILE

A stunningly beautiful species that is occasionally offered for sale in the hobby. It is a flattened starfish, with short rounded ends to the arms.

WHAT size?
Grows to about 15cm across the arms.

WHAT does it eat?
Prefers encrusting organisms, but its exact aquarium requirements are unknown. It may browse on sponges, bryozoans, bacteria and similar organisms, meaning that it is almost impossible to supply a specific diet. Best stocked into a mature aquarium that has had a chance to develop sustainable populations of potential food sources.

WHERE is it from?
Pacific. Its range falls outside tropical zones.

WHAT does it cost?
★★★☆☆ ★★★★☆
Expensive due to scarcity and its stunning appearance.

▶ *The appeal of this animal is easy to see, but if water temperatures are not carefully controlled, this subtropical species is unlikely to survive for long.*

HOW many in one tank?
Best kept singly.

HOW compatible with other invertebrates?
May attack sponges, but unlikely to present a threat to other aquarium invertebrates.

WILL fish pose a threat?
Few fish will bother this starfish. Angelfish and butterflyfish might attempt to nip at its tube feet.

WILL it threaten fish?
No, it offers no threat to any fish in the aquarium.

WHAT to watch out for?
This is a subtropical species that may survive long term in a reef aquarium kept at a temperature of about 24°C, but it prefers cooler conditions. This starfish needs careful

SIMILAR SPECIES
Not likely to be confused with any other commonly available starfish, but *Fromia monilis* has a similar red blotched appearance. However, the flattened rigid shape of the biscuit star is characteristic and a useful aid in its identification.

acclimatisation and should be left in the dealer's tank for a prolonged period before you buy it. This gives it time to recover from the rigours of shipping and might offer you some insights into suitable substitute feeds.

WILL it reproduce in an aquarium?
No.

Protoreaster lincki

Crimson knobbed starfish

A large and hardy starfish that is collected from seagrass beds in its native range. Its coloration may advertise its unpalatability to many fish species.

WHAT size?
Grows to about 20cm across the arms.

WHAT does it eat?
Will eat most foods in the aquarium, but be sure to provide a varied diet, including chopped shellfish, mysis and similar meaty foods. As it is known to feed on sponges, at least to some extent, those diets formulated for marine angelfish may be useful in catering for its long-term nutritional requirements.

WHERE is it from?
Tropical Indo-Pacific.

WHAT does it cost?
★★★☆☆
Its large size (for a starfish) makes it more expensive than many others, even though it is readily available.

SIMILAR SPECIES
Closely related to the chocolate chip starfish *(Protoreaster nodosus)*, but that species has dark brown raised nodules on its dorsal surface.

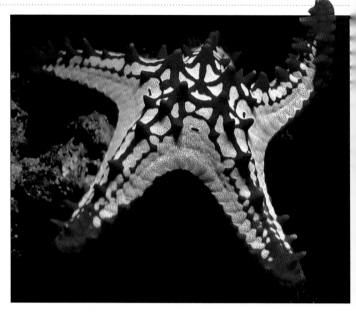

▲ *Given a live rock-based aquarium and a varied diet, this starfish will do well in peaceful fish-only systems.*

HOW many in one tank?
Can be kept in numbers, but the onus is on the aquarist to provide sufficient food for each one.

HOW compatible with other invertebrates?
Will consume a variety of sessile invertebrates, including tridacnid clams, sponges, anemones and, sometimes, zoanthid button polyps. Some aquarists report isolated incidents of soft and stony corals being targeted.

WILL fish pose a threat?
Few fish will bother this starfish.

Angelfish and butterflyfish might attempt to nip at its tube feet.

WILL it threaten fish?
No. it offers no threat to any fish in the aquarium, although it may scavenge dead specimens.

WHAT to watch out for?
Only select animals that have been settled into the dealer's tank for a week or more. Although hardier than many starfish commonly offered for sale, it is sensitive to sudden changes in water quality.

WILL it reproduce in an aquarium?
No. It may spawn but this is unlikely to be successful.

Protoreaster nodosus

Chocolate chip starfish

PROFILE

Of all the starfish commonly imported for the aquarium hobby, this species is arguably the easiest to maintain in the home aquarium. It is hardy, long-lived, omnivorous and inexpensive, yet it has fallen out of favour with many aquarists due to its unpredictable behaviour in the aquarium. In some instances it behaves reasonably well, in others extremely badly!

WHAT size?
Grows to about 15cm across the arms.

WHAT does it eat?
Will consume almost anything in the aquarium, from uneaten food intended for fish to certain sessile invertebrates. Provide regular offerings of meaty foods, including chopped shellfish and mysis. Consumes algae but is likely to be fussy in its preferences for green foodstuffs.

WHERE is it from?
Tropical Indo-Pacific.

WHAT does it cost?
★☆☆☆☆ ★★☆☆☆
Inexpensive.

▶ *This starfish's common name derives from the dark brown protruberences along its upper surface. It proves hardy once it has recovered from the rigours of collection and shipping.*

HOW many in one tank?
Can be kept in numbers where space and food allow.

HOW compatible with other invertebrates?
Will consume a variety of sessile invertebrates, including tridacnid clams, sponges, anemones and, sometimes, zoanthid button polyps.

WILL fish pose a threat?
This starfish is sometimes kept in live rock-based, fish-only aquariums. Avoid keeping it with species such as triggerfish that may take investigatory nibbles at its arms. Angelfish and butterflyfish may target the tube feet.

WILL it threaten fish?
No, it offers no threats to any fish in the aquarium, although it may scavenge dead specimens.

WHAT to watch out for?
Avoid specimens showing any signs of damage on the arms

SIMILAR SPECIES
Many species of starfish have a grey-brown base colour with raised pointed nodules over the dorsal surface. However, these projections have characteristic colours. In the crimson knobbed starfish (*Protoreaster lincki*), the common name explains the animal's appearance.

or body. In common with all starfish, this species is sensitive to sudden changes in water quality and/or chemistry, and can literally disintegrate over a short period of time. Select specimens that have been settled for a while in the dealer's aquarium.

WILL it reproduce in an aquarium?
No. It may spawn but this is unlikely to be successful.

Astroboa sp.

Basket starfish

PROFILE

Basket starfish are a fascinating and beautiful group of starfish in which the arms have become so highly branched that when extended they resemble a net or woven basket. They are similar in their demands to feather starfish, but unless you can provide the utmost care and attention, leave them in the dealer's tank.

WHAT size?

Most aquarium specimens are fairly small when first acquired, perhaps 15cm across the arms. Maximum size will depend on species, but some can reach over 1m across.

WHAT does it eat?

Offer regular daily feeds with larger plankton substitutes. Frozen marine copepods, brineshrimp, mysis and gammarids are all suitable, as are frozen and dried invertebrate foods. Phytoplankton is too small to be eaten, but may tempt the animal to spread its arms. Basket stars may only emerge to feed at night, although in time they can be trained to begin feeding by day.

WHERE is it from?

Circumtropical.

WHAT does it cost?

★★☆☆☆   ★★★☆☆
Sometimes free on a piece of coral or other invertebrate.

HOW many in one tank?

Due to its specialised feeding requirements, this animal is best kept singly, unless the aquarist can confidently provide for the long-term needs of two or more.

HOW compatible with other invertebrates?

Will not harm other invertebrates, even if it decides to use one as a vantage point from which to feed. Shrimps, hermit and true crabs may raid a basket star's arms at feeding time and can damage it.

WILL fish pose a threat?

Most fish generally considered to be reef-safe will ignore basket starfish, although some may attempt to remove food from the arms. This behaviour may damage the invertebrate.

WILL it threaten fish?

No. Basket starfish do not predate or harm fish in any way.

WHAT to watch out for?

Specimens in a dealer's aquarium are likely to be curled up, sometimes to the extent that they resemble a small ball of worms, rather than an echinoderm. This is entirely natural. Check for signs of discoloration, damage to the arms or any deposits or residue around the area where the animal is resting, as these can indicate bacterial infections and/or tissue damage. Acclimatise over an extended period.

WILL it reproduce in an aquarium?

Unlikely, although not impossible.

▼ *Basket starfish are beautiful animals that should only be purchased by aquarists who are fully aware of their aquarium demands.*

Comanthina sp.

Yellow-tipped green feather starfish

PROFILE

This feather starfish is only sporadically available, yet many aquarists report success with it, sometimes without employing any special husbandry techniques. However, this is no excuse for neglecting to provide the animal's particular needs.

WHAT size?
Grows to about 25cm across the arms.

WHAT does it eat?
Probably feeds on a greater diversity of particulate material than other regularly imported feather starfish. Offer a variety of particulate live and frozen plankton, plus a selection of the many formula foods now available for filter-feeding invertebrates. This species may only emerge under actinic or moonlight illumination and must be offered food when the arms are extended. With perseverance, it is possible to train it to come out during the day.

WHERE is it from?
Tropical Indo-Pacific. Specimens are chiefly collected from Indonesia and the Philippines.

WHAT does it cost?
★★☆☆☆ ★★★☆☆
Medium priced.

◀ There are many species of feather starfish that are imported for the aquarium trade with varying degrees of hardiness. Aquarists should be aware of the demands of a particular individual before purchase.

HOW many in one tank?
Can kept in numbers, but adding more food to the aquarium means monitoring the water quality carefully.

HOW compatible with other invertebrates?
Should not present any problems to invertebrates, whether sessile or free-living.

WILL fish pose a threat?
Certain fish may decide to nibble at the arms of this feather starfish, some to the extent that they may kill it. It is difficult to identify the guilty species, as almost any browsing fish may take an active interest in molesting the crinoid, but angelfish and dwarf angelfish are likely to be the main culprits.

WILL it threaten fish?
Will not harm any species of fish.

WHAT to watch out for?
When first encountered, specimens are often retracted. Their arms curl up, forming a roughly spherical mass around the central hub. This is entirely normal during the aquarium's daylight period. Check for signs of arm damage and avoid any specimens with discoloured patches anywhere on the arms or body. Acclimatise them slowly and carefully.

WILL it reproduce in an aquarium?
Unlikely, but not impossible. Eggs are brooded in the arms.

SIMILAR SPECIES

The yellow-gold tips to the thin radiating branches of the green arms are a good clue to the correct identification of this species. There are other species with a similar appearance, but few seem to be imported for the aquarium trade.

Himerometra robustipinna

Red feather starfish

This scarlet species is arguably the most stunning of all of the feather starfish and tempts many aquarists who lack the knowledge to keep it healthy. Do not buy it unless you are willing to dedicate the time and energy needed to provide for its long-term needs.

WHAT size?
Grows to about 25cm across the arms.

WHAT does it eat?
Providing a diet of sufficient diversity, nutritional content and in adequate quantities is the greatest challenge facing potential keepers of this exquisite animal. Offer phytoplankton, microplankton 'marine snow' and any other very small particulate feeds. As the red feather starfish is primarily nocturnal, it may be necessary to drip food into the aquarium at night for the best results.

WHERE is it from?
Tropical Indo-Pacific. Specimens are chiefly collected from Indonesia and the Philippines.

WHAT does it cost?
★★★☆☆ ★★★★☆
Can be expensive, depending on origin and whether it has been imported directly or through a wholesaler.

◀ *The undeniably beautiful red feather starfish should be left to settle before purchase. Always check the area around this animal for signs of the disintegration that occurs in ailing specimens.*

HOW many in one tank?
Best kept singly.

HOW compatible with other invertebrates?
Should not present any problems to invertebrates, whether sessile or free-living.

WILL fish pose a threat?
Certain fish may decide to nibble at the arms of this feather starfish, some to the extent that they may kill it. It is difficult to identify the guilty species, as almost any browsing fish may take an active interest in molesting the crinoid, but angelfish and dwarf angelfish are likely to be the main culprits.

WILL it threaten fish?
Will not harm any species of fish.

WHAT to watch out for?
Unhealthy specimens often begin to register their distress by dropping arms or fragments of arms. Always check around a potential purchase for signs of this occurring. Many will hide when the aquarium lights are on, emerging at night to spread their arms and feed. As with other echinoderms, it is a good idea to acclimatise specimens over a few hours, gradually adding water from the aquarium to their transportation bag.

WILL it reproduce in an aquarium?
Unlikely, although not impossible.

SIMILAR SPECIES
This species of feather starfish is unlikely to be confused with any other, at least of those that make their way into the aquarium trade. There is a very closely related brown species with almost identical aquarium requirements.

Ophiarachna incrassata

Green brittle starfish

PROFILE

Many aquarists value the scavenging behaviour of this species and it is commonly kept, but its usefulness is outweighed by its potential to cause harm. It targets small fish by raising its body above the substrate, creating a refuge beneath which its victims may be tempted to rest. It then coils its arms by twisting the body and closing the trap or ensnaring the fish on the sharp lateral spines at the edges of the arms.

WHAT size?

Grows to about 45cm across the arms.

WHAT does it eat?

This animal is happy to scavenge uneaten fish food or will accept meaty foods of quite a large size. The fact that it will readily accept fish is a real clue to its true nature.

WHERE is it from?

Tropical Indo-Pacific.

WHAT does it cost?

★★☆☆☆
Relatively inexpensive; larger specimens can command a higher price.

HOW many in one tank?

Can be kept in groups where small fish are absent.

HOW compatible with other invertebrates?

Will eat free-living invertebrates, such as shrimp, crabs and similar organisms.

WILL fish pose a threat?

Fish ignore this brittle starfish.

WILL it threaten fish?

Yes. Potentially lethal to a number of fish species. Small species, fish that live close to the substrate or that spend the night there are particularly vulnerable. This brittle starfish should be kept only with large fish.

WHAT to watch out for?

There are few documented problems with this species. It is even safe to buy an individual with missing legs in the knowledge that these

SIMILAR SPECIES

Could be confused with other members of the genus or species from the genera *Ophiomastrix* (which can also trap small fish, albeit without the ruthless efficiency of the green brittle starfish). Could also be confused with *Ophionereis* and *Ophiocoma*.

will regrow. However, it is worth observing a specimen for a period, perhaps even introducing some food to ensure that it is mobile and has no obvious damage.

WILL it reproduce in an aquarium?

This is one of the few echinoderms that has been reported to spawn regularly in the home aquarium. The young will grow to adulthood.

◀ *Interesting, inexpensive and useful are just three words that describe this animal. Highly predatory are two more. Do not keep it with small fish.*

169

Ophioderma cf. *appressum*

Marbled serpent starfish

PROFILE

A beautiful species available at small sizes (less than 10cm across the arms). It has highly variable patterning, meaning that almost every specimen is unique in appearance. These highly mobile serpent starfish can move quickly to evade predators or when in search of food. The body can be brown, green or cream or any combination of these colours.

WHAT size?
Grows to about 30cm across the arms.

WHAT does it eat?
Happy to scavenge almost anything in the aquarium, including fish faeces and uneaten food intended for fish or shrimp. As it is largely self-sufficient in terms of locating food, it is unlikely ever to need target-feeding. However, you can provide regular offerings of any meaty items, such as pieces of shellfish, mysis or flaked and granular foods.

WHERE is it from?
Tropical Atlantic – Eastern and Western.

WHAT does it cost?
★☆☆☆☆   ★★☆☆☆
Inexpensive. The price range shown here is for small specimens to the largest available.

HOW many in one tank?
Can be kept in numbers, especially when small. The animals have large appetites, so provide plenty of food as they grow.

HOW compatible with other invertebrates?
Should not present any problems to invertebrates, whether sessile or free-living.

▲ *This benign creature plays a highly useful role in the aquarium, but many aquarists stock it for its looks alone.*

WILL fish pose a threat?
Inquisitive fish may occasionally peck at specimens of this serpent starfish, but such behaviour seldom persists.

WILL it threaten fish?
Will not harm any species of fish.

WHAT to watch out for?
Choose specimens with vibrant colours. Marbled serpent starfish are as vulnerable to careless acclimatisation and the stresses of transport as other starfish. Avoid any with patches of discoloration; these become evident as bacterial infections take hold. Specimens missing legs will grow new ones, and are safe to buy, as long as they are otherwise healthy.

WILL it reproduce in an aquarium?
Unlikely, although not impossible.

SIMILAR SPECIES
There are several species of serpent starfish with a brown-olive base colour and prominent dark bands encircling the arms. One is *Ophiolepis superba*, commonly imported from Indonesia. Although useful as a scavenger in its own right, it is not as easy on the eye as its relative from the tropical Atlantic.

Ophioderma squamossimus

Red serpent starfish

PROFILE

This potentially large and highly mobile species is one of the most stunning of all the brittle starfish. Serpent starfish generally lack the lateral arm bristles seen in brittle starfish, even though they are closely related. Apparently, they are also less likely to break or drop their arms when stressed or handled.

WHAT size?
Grows to about 45cm across the arms.

WHAT does it eat?
Happy to scavenge almost anything in the aquarium, including fish faeces and uneaten food intended for fish or shrimp. Provide regular offerings of any meaty items, such as pieces of shellfish, mysis or flaked and granular foods.

WHERE is it from?
Tropical Atlantic – the Caribbean. Often referred to as the Bahamas serpent starfish.

WHAT does it cost?
★★★☆☆ ★★★★★
Price will depend on the location of the end market.

▶ *This highly sought-after species of serpent starfish commands a high price that it more than justifies.*

HOW many in one tank?
Due to its large size it is best kept singly, unless the aquarium is 600 litres or more.

HOW compatible with other invertebrates?
Has the potential to consume slow-moving shrimps or crabs, but this rarely happens as most serpent starfish coexist peacefully with other invertebrates.

WILL fish pose a threat?
Fish ignore this brittle starfish.

WILL it threaten fish?
Might scavenge dead or dying fish, but is extremely unlikely to catch and kill them.

SIMILAR SPECIES
Not likely to be confused with any other species of serpent starfish due to its crimson coloration. The ruby brittle star (*Ophioderma rubicundum*) is closely related and also has red pigment, but this is punctuated by cream bands that are particularly evident on the arms.

Its efficiency at consuming carcasses, and the tell-tale swollen body that results from a hearty meal, probably explains why it is often blamed for otherwise mysterious fish deaths.

WHAT to watch out for?
Avoid specimens with patches of discoloration, particularly on the legs. Red serpent starfish are susceptible to bacterial infections that result from poor acclimatisation or a deterioration in water quality experienced when they are shipped. If possible, try to observe a specimen feeding.

WILL it reproduce in an aquarium?
Unlikely, although not impossible.

Ophiomastrix spp.

Black brittle starfish

PROFILE

A readily available species of brittle starfish stocked for its scavenging behaviour. Although unlikely to win any beauty contests, it justifies its inclusion in most aquariums on account of this role and is a fascinating beast to behold.

WHAT size?
Grows to about 45cm across the arms.

WHAT does it eat?
This animal is happy to scavenge uneaten fish food and fish waste or will accept meaty foods of quite a large size. It also consumes dead fish, with the result that it is sometimes falsely accused of causing a fish's death, rather than being credited with cleaning up the crime scene.

WHERE is it from?
Circumtropical.

WHAT does it cost?
★☆☆☆☆ ★★☆☆☆
Inexpensive. Price depends on size.

▶ *Provided the buyer can be confident that their potential purchase is not one of the brittle starfish known to predate fish, several species of these echinoderms are suitable for marine aquariums.*

HOW many in one tank?
Can be kept in numbers. The animals have large appetites, so provide plenty of food.

HOW compatible with other invertebrates?
Should not harm ornamental invertebrates such as shrimps or crabs. However, brittle starfish can be clumsy animals with the potential to knock over poorly anchored invertebrates.

WILL fish pose a threat?
Fish that are generally considered to be reef-safe usually ignore this brittle starfish, but take care when attempting to house it in live rock-based, fish-only aquariums that are home to boisterous fish with inquisitive natures.

WILL it threaten fish?
No. Unlike the green brittle starfish, this species does not attempt to trap and kill fish.

WHAT to watch out for?
Make sure that specimens do not show any signs of discoloration. Avoid any with white patches, which can indicate bacterial infections or necrosis.

WILL it reproduce in an aquarium?
Unlikely, although not impossible.

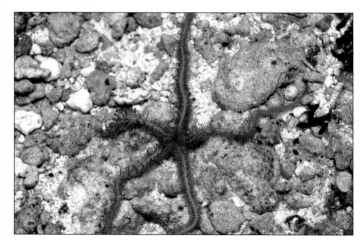

Asthenosoma spp.

Fire urchin

PROFILE

The good looks of this irregular import will appeal to many aquarists, but in the long term it is really a beast for the urchin enthusiast. Potentially, it is very dangerous for aquarists to come into contact with it. Although it appears to adapt well to aquarium life, problems manifest themselves during the transfer of the animal from one aquarium to another and in avoiding injury both to the aquarist and the urchin.

WHAT size?
Large specimens reach 15cm or so in diameter.

WHAT does it eat?
Primarily herbivorous, but will appreciate regular offerings of dried algae if and when it depletes the naturally occurring forms in the aquarium. It may occasionally scavenge uneaten food intended for fish.

WHERE is it from?
Tropical Indo-Pacific.

WHAT does it cost?
★★★☆☆　★★★★☆
This rare import sometimes commands a high price. Do not buy a specimen unless you are confident that you can provide for its long-term needs and simultaneously bear in mind your own safety.

▲ This echinoderm is both beauty and beast, due to the defensive venom housed in its spines.

HOW many in one tank?
Can be kept with others of the same species.

HOW compatible with other invertebrates?
Has the potential to inject venom into anything that comes too close. This could include sessile invertebrates as it moves around the aquarium, but this is very unlikely.

WILL fish pose a threat?
Most fish are well aware of the dangers this sea urchin presents to them and will steer well clear of it.

WILL it threaten fish?
Unlikely, but not impossible in the event of an attack by a fish.

WHAT to watch out for?
This urchin is capable of injecting a very nasty venom into its victims and causing massive pain to careless or plain unlucky handlers. Always use thick gloves or nets to move this animal, and locate its position in the aquarium before you begin any maintenance work inside. Occasionally, this urchin may play host to probably commensal shrimp from the genus *Periclimenes*.

WILL it reproduce in an aquarium?
Potentially possible, but as few people maintain this urchin in an aquarium, there are few reports of spawning events.

SIMILAR SPECIES

Bears a superficial resemblance to *Toxopneustes* spp. sea urchins, but the spines of the fire urchin have obvious swellings below their tips that house the venom. Almost all species are very colourful; the Red Sea *A. varium* is blood red-and-orange, with creamy-white venom sacs. One of the more commonly available fire urchin species is *A. ijimai*, which has venomous purple spines and a rainbow of other colours on the test (body) and spines.

Diadema setosum

Longspine sea urchin

A stunning species of sea urchin, often available at a small size that tempts the unsuspecting aquarist to buy it. However, it grows rapidly and can quickly become too large for many aquariums. Its dimensions are only one of the potential drawbacks in its long-term maintenance; it can deliver a painful injury if not treated with caution.

WHAT size?
Can reach over 40cm across the spines. The body is much smaller, usually less than half of this, where the spines have not become damaged.

WHAT does it eat?
Feeds on calcareous forms of algae and leafy macroalgae, as well as filamentous forms. Dried algae is useful. In order to grow, it needs plenty of calcium in its diet, so often strips live rock of its colourful algae. Where provided, it will rasp on cuttlebone as an extra source of this essential element.

WHERE is it from?
Tropical Indo-Pacific.

WHAT does it cost?
★☆☆☆☆ ★★★☆☆
Price depends on size and origin.

HOW many in one tank?
Can be kept in numbers.

HOW compatible with other invertebrates?
Unlikely to consume any invertebrates or attack them directly, but can knock them over or damage them as it moves by.

WILL fish pose a threat?
Very few fish bother with this urchin. A species of clingfish (*Diademichthys lineatus*), which is occasionally imported for the aquarium trade, is known to feed on the tube feet of this animal.

WILL it threaten fish?
This urchin has the potential to harm fish with its spines, but this is extremely rare. Many fish, such as the aquarium favourite Banggai cardinal (*Pterapogon kauderni*), actually use the spines as a refuge or nursery.

WHAT to watch out for?
This urchin needs time to acclimatise. Do not buy one unless you really want to keep it and are prepared to forgive its indiscretions. Capable of inflicting very painful wounds if handled carelessly. The hollow spines contain venom. These, coupled with the backward-pointing barbs, inflict injuries that can cause discomfort for weeks or months. Treat the urchin with respect, especially when large.

WILL it reproduce in an aquarium?
Commonly reported spawning in the aquarium, but this does not result in the production of offspring.

◀ *The prominent protruberance on the upper surface, commonly assumed to be an eye, is actually the animal's anal sac.*

Heterocentrotus mammillatus

Slate pencil urchin

PROFILE

A beautiful species of urchin, named for the large robust spines that are sometimes used in decorative works or jewellery. However, these are only one of the urchin's two types of spine. The other spines are very short and capable of inflicting painful injuries on the unwary aquarist.

WHAT size?

The long spines of this urchin can give it an overall diameter in excess of 20cm.

WHAT does it eat?

Offer dried seaweed, particularly nori. Will consume calcareous algae and also scavenge uneaten food intended for fish. Some pencil urchins have been reported trapping and eating fish.

WHERE is it from?

Tropical Indo-Pacific.

WHAT does it cost?

★★☆☆☆   ★★★☆☆

Price depends on size. Very small specimens measuring less than 4cm are occasionally available, but they grow very quickly.

HOW many in one tank?

Can be kept in numbers. Provide plenty of food.

HOW compatible with other invertebrates?

Very likely to knock over any sessile invertebrate that is not securely attached to the reef rockwork. This is a deceptively strong animal that will soon find out how competently the aquarist has secured the aquarium decor into place.

WILL fish pose a threat?

Few fish present any problems to this urchin. Indeed, given the potential risks to the reef aquarium of stocking it, this animal is probably best suited to a live rock-based, fish-only system.

WILL it threaten fish?

Some species of pencil urchin have been reported killing fish with the aid of their spines. This is not a recognised trait of *H. mammillatus*.

WHAT to watch out for?

Avoid specimens that appear listless or have drooping spines, as they may not have fully recovered from the rigours of collection and shipping. Healthy individuals very quickly

▲ *Thick, solid spines are characteristic of the pencil urchins.*

strip colourful algae from the rockwork, so provide plenty of algae to meet the urchin's long-term nutritional requirements.

WILL it reproduce in an aquarium?

Will sometimes spawn in the aquarium, releasing its gametes directly into the water column. Wild specimens breed only once a year.

SIMILAR SPECIES

Many species of pencil urchin may be offered for sale on a fairly regular basis. *H. mammillatus* has distinctive spines that are triangular in cross-section. The Caribbean species *Eucidaris tribuloides* is known to feed on encrusting organisms and even bivalve molluscs.

Lytechinus variegatus

Pincushion urchin

PROFILE

An interesting species, with a number of colour varieties, regularly imported for the aquarium trade. It is often bought to help control nuisance algae in the reef aquarium by aquarists who have not researched it carefully and are unaware of the potential problems it may cause. Using its tube feet it picks up rubble, shells and other material to cover its spines as camouflage.

◀ *Pincushion urchins are highly variable in coloration, but always have relatively short spines that may be obscured by the assorted debris this animal picks up as it moves around the aquarium.*

WHAT size?

Specimens offered for sale can measure 2.5-5cm, which belies their large final size; they can reach over 20cm across the test. The spines are short and do not increase the overall diameter of the animal significantly.

WHAT does it eat?

Feeds on a variety of algae types. When it has consumed much of the naturally occurring greens in the aquarium, offer it dried green seaweed. Providing cuttlebone can help to satisfy the animal's demand for calcium, which it needs in order to grow.

WHERE is it from?

Tropical Atlantic.

WHAT does it cost?

★☆☆☆☆   ★★☆☆☆
Price depends on size.

HOW many in one tank?

Can be kept in groups, but will severely affect calcareous algal growth.

HOW compatible with other invertebrates?

Very likely to knock over any sessile invertebrate that is not securely attached to the reef rockwork. Will pick up small specimens to assist in camouflage, often dropping them behind rockwork.

WILL fish pose a threat?

Few fish generally considered to be reef-safe will bother the pincushion urchin. Some may show some interest in the assorted jetsam that the urchin covers itself in, but should not attack the animal itself.

WILL it threaten fish?

No, harmless towards fish.

WHAT to watch out for?

A potentially destructive animal that can best be described as clumsy, rather than predatory or anything more sinister. It picks things up and knocks them over, often to the consternation of the aquarist who originally bought it to do a beneficial job in the aquarium.

WILL it reproduce in an aquarium?

Will spawn but impossible to sex. Sperm and eggs are released into the water column, where they are usually removed quickly by filters or protein skimming.

SIMILAR SPECIES

Few urchins commonly sold in the aquarium trade are likely to be confused with the pincushion. Its short spines, arranged regularly over the upper surface of the animal, are quite distinctive.

Mespilia globulus

Tuxedo, or 'spineless', urchin

PROFILE

A small species of sea urchin and generally considered to be one of the most suitable for the reef aquarium. It tends not to decimate calcareous algae and, being smaller than some urchins, is less likely to knock over sessile invertebrates. The spines are short and bristle-like, punctuated by areas of smooth, blue-pigmented shell.

WHAT size?
May reach 6cm in diameter, including spines.

WHAT does it eat?
Feeds almost exclusively on algae. Will consume filamentous forms and several other species of green algae. Offer dried algae when the aquarium's naturally occurring supplies have been exhausted.

WHERE is it from?
Tropical Indo-Pacific.

WHAT does it cost?
★☆☆☆☆ ★★☆☆☆
Price depends on size and origin.

SIMILAR SPECIES
The tuxedo urchin is unlikely to be confused with any other commonly available urchins.

HOW many in one tank?
The least problematic sea urchin commonly available in the trade. It can be housed in numbers where desired.

HOW compatible with other invertebrates?
Reported to consume soft corals from the genus *Paralemnalia*. However, such instances are rare and in general this urchin is reef-safe.

WILL fish pose a threat?
Few fish wish to tackle any species of sea urchin. Certain triggerfish have evolved a technique for directing jets of water from their mouths at the base of the invertebrate, turning it over and exposing its vulnerable underside. However, such triggerfish are not usually

▼ *Tuxedo urchins often pick up pieces of debris around the aquarium in order to disguise themselves.*

kept in the reef aquarium. Triggerfish that are kept, such as members of the genus *Xanthichthys,* do not employ such methods.

WILL it threaten fish?
No.

WHAT to watch out for?
Acclimatise the urchin over an extended period of a couple of hours or so, gradually mixing the water throughout this time. Tuxedo urchins will pick up fragments of almost anything in the aquarium, from small polyps to algae, which endow them with a form of camouflage.

WILL it reproduce in an aquarium?
Many species of sea urchin will release sperm or eggs in the home aquarium, but as they tend to be kept singly, such events do not lead to the production of any offspring.

Metalia spp.

Heart urchin

PROFILE

The heart urchins, or sea potatoes, are irregular imports that could conceivably find considerable popularity with marine aquarists for the potential benefits they would bring to the system. They are unusual in being one of the few groups of urchins that burrow into soft substrates. They have several anatomical adaptations that enable this to happen, not least the abandonment of the roughly spherical shape of their relatives.

WHAT size?
Grows to about 7cm long, not including the thin, hairlike spines.

WHAT does it eat?
The heart urchin is a scavenger and detritivore. It may consume algae and other plant material in the form of sea grass. Offer dried forms of algae. May thrive in a caulerpa-rich, macroalgae aquarium as part of a larger reef system, providing there is sufficient substrate for burrowing.

WHERE is it from?
Tropical Indo-Pacific.

WHAT does it cost?
★★☆☆☆
Relatively inexpensive.

HOW many in one tank?
Can be kept with others of the same species. Upper limits will depend on the amount of substrate present and food availability.

HOW compatible with other invertebrates?
Its burrowing behaviour has the potential to undermine rockwork, but this is not a strong animal and it will usually take the easiest route, rather than plough a more difficult furrow.

WILL fish pose a threat?
With its relatively thin test and lack of any substantial defensive spines (that would otherwise impede its burrowing existence), the heart urchin is potentially vulnerable to attack from large or inquisitive fish. However, few seem to take any notice of it when it emerges from the

substrate, so it can be stocked with a reasonable degree of confidence.

WILL it threaten fish?
No. This urchin is entirely harmless towards fish.

WHAT to watch out for?
Ideally, observe specimens burrowing before you buy them. They should have roughly symmetrical patterns of hairlike spines. Patches where the test can be seen and hairs/spines are absent may signify a damaged or ailing individual.

WILL it reproduce in an aquarium?
Very unlikely, but cannot be ruled out.

◄ *The feeding and burrowing behaviour of the heart urchins can help to keep the sand aerated and relatively free of detritus.*

Toxopneustes spp.
Toxic sea urchin

PROFILE

Given its stunning appearance and sporadic availability, aquarists may be forgiven for thinking that this is an extremely desirable species of sea urchin. However, its defences are formidable and although it will not cause the potentially large puncture wounds of the long-spined sea urchin, it will certainly make those who come into contact with it aware of their mistake. If the victim is particularly sensitive to the poisons of this animal, the results of being stung are potentially lethal.

WHAT size?
Large specimens reach 15cm or so in diameter.

WHAT does it eat?
Readily accepts small chunks of almost any shellfish. Mysis and brineshrimp are also suitable. Offer these about once a week. At other times, the urchin consumes algae of various descriptions and welcomes the addition of dried algae to the aquarium.

WHERE is it from?
Tropical Indo-Pacific.

WHAT does it cost?
★★☆☆ ★★★★☆
Price depends largely on size.

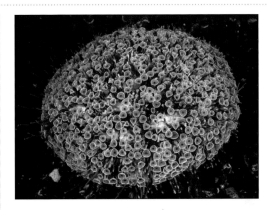

◀ *Toxic urchins are blessed with a stunning appearance that belies their powerful defences. Reactions to being stung can be severe, so always treat the animal with great respect.*

HOW many in one tank?
Can be kept with others of the same species.

HOW compatible with other invertebrates?
Has the potential to 'sting' anything that comes too close. This could include sessile invertebrates as it moves around the aquarium.

WILL fish pose a threat?
Most fish are well aware of the dangers that this sea urchin presents to them and will steer well clear of it.

WILL it threaten fish?
Unlikely but not impossible if attacked by a fish.

SIMILAR SPECIES
Unlikely to be confused with any other sea urchins due to their prominent pedicellariae.

WHAT to watch out for?
This urchin does not sting, as such. In common with most other sea urchins and some starfish, it has specialised structures called pedicellariae that resemble three-fingered robotic claws. Some species use the claws to pass food across their surface or to hold onto various items for the purposes of camouflage. In *Toxopneustes*, each one can pierce the skin. The base of the 'claw' holds venom that can be 'injected' into the victim. Handle the animal with care; always wear thick rubber gloves and/or use a net. When cleaning the aquarium, take care not to brush the urchin accidentally.

WILL it reproduce in an aquarium?
Potentially possible, but as few people maintain this urchin in the aquarium, there are few reports of spawning events.

Colochirus robustus

Yellow sea cucumber

PROFILE

This beautiful animal is
one of the most commonly
imported sea cucumbers for
the aquarium trade. It is also
likely to be one of the safest
of its kind to introduce to
the aquarium. The maximum
size of this sea cucumber
would appear to make it
suitable for including in a
nanoreef aquarium, but given
its potential to release toxic
compounds into the water, do
not introduce it into systems of
less than 100 litres.

WHAT size?
5-6cm body length.

WHAT does it eat?
Feeds on particulate material
extracted from the water
column. Its branching
arms return to its mouth
one at a time in sequence.
Offer rotifer-based formula
foods, phytoplankton and
zooplankton daily for the
best results.

WHERE is it from?
Tropical Indo-Pacific.

WHAT does it cost?
★☆☆☆☆   ★★☆☆☆
Inexpensive.

HOW many in one tank?
Can be kept with
others of the same
species or even
close relatives,
providing there is
sufficient food.

HOW compatible with other invertebrates?
Sometimes falls
victim to aggressive crustaceans,
such as crabs or hermit crabs.

WILL fish pose a threat?
Occasionally fish will peck at
the feeding arms. Whether
this is through aggression,
inquisitiveness or simply to
remove captured food from the
feeding arms is not clear.

WILL it threaten fish?
Not reported to cause any

SIMILAR SPECIES
There are several similar
species of sea cucumber,
such as *Pentacta* sp. from
the Indo-Pacific, with an
almost identical overall
body shape. Aquarists
should not assume that
their appearance guarantees
their suitability for the
invertebrate aquarium,
as many are capable of
causing fish deaths through
the release of defensive
chemicals.

▲ *Colochirus robustus is widely
perceived to be the safest sea
cucumber to stock in an aquarium
containing fish. Be sure to cover
pump inlets and drains to prevent
damage to the animal.*

fish deaths in the aquarium
through the release of defensive
chemicals. However, this cannot
be ruled out, particularly in small
aquariums.

WHAT to watch out for?
Try to identify specimens that
have their arms extended and are
in the process of feeding. Check
the body surface for areas of
discoloration and avoid individuals
where these are present.

WILL it reproduce in an aquarium?
Asexual reproduction is fairly
common in healthy individuals.
They split in two, roughly halfway
along the body, and each grows
its missing parts.

Holothuria edulis

Pink hot dog sea cucumber

PROFILE

Although it is one of the least attractive of the sea cucumbers regularly imported for the aquarium trade, this species nevertheless has its fans. It processes sand in a similar way to earthworms, sucking up the substrate, digesting its nutritional contents and excreting the indigestible medium in cylinder-shaped pellets.

WHAT size?

Grows to around 30cm total body length.

WHAT does it eat?

Bacteria, detritus and small interstitial organisms living in sand. This diet can be supplemented (but not easily) by adding particulate foods, such as 'marine snow', that are allowed to settle onto the substrate.

WHERE is it from?

Tropical Indo-Pacific.

WHAT does it cost?

★★☆☆☆
Relatively inexpensive.

HOW many in one tank?

Best kept singly. This efficient feeder will rapidly exhaust naturally occurring food in the reef aquarium.

HOW compatible with other invertebrates?

May be bothered by large hermit crabs or other crustaceans. One species of true crab is known to live commensally inside the anus of similar sea cucumbers.

WILL fish pose a threat?

Few fish seem to bother with this sea cucumber.

WILL it threaten fish?

Can release the potent toxin holothurin when stressed or even for no obvious reason. This fast-acting agent can be deadly to fish.

WHAT to watch out for?

May coexist peacefully with its

SIMILAR SPECIES

Several members of the genus *Holothuria* are commonly imported for the aquarium trade. *H. hilla* can be identified by its brown body colour and the lighter 'thorns' that cover the body. The black species *H. atra* uses short tentacles to mop up detritus from the surface of the sand. All should only be stocked in the knowledge that their potential benefits can be outweighed by the possibility that they may cause severe harm to fish.

tankmates for years and then for no apparent reason suddenly squirt holothurin into the water, resulting in the death of the fish stocks. Members of this genus can also voluntarily eviscerate themselves in times of stress.

WILL it reproduce in an aquarium?

Some species from the genus *Holothuria* are known to reproduce asexually. It is likely that most will do so if they are adequately fed.

◀ *The short feeding tentacles of this animal are retracted when it is at rest. It ingests sand particles and excretes sausage-like faecal pellets.*

Pseudocolochirus axiologus

Sea apple

Its unusual and stunning appearance means that the sea apple satisfies most aquarists' longing for a novel and striking animal for their reef aquarium. However, it has the potential to be extremely destructive, so do not buy it unless you intend to keep it in a specialised system containing no fish.

WHAT size?
Grows to around 15cm body length with the arms retracted. When stressed, it can swell to almost twice this size. What benefit this gives the animal is unclear.

WHAT does it eat?
Feeds on particulate material extracted from the water column. Its branching arms return to its mouth one at a time in sequence. Offer rotifer-based formula foods, phytoplankton and zooplankton daily for the best results.

WHERE is it from?
Tropical Indo-Pacific.

WHAT does it cost?
★★☆☆☆ ★★★☆☆
Medium priced.

▲ *Many aquarists live to regret stocking the sea apple into their aquariums.*

HOW many in one tank?
Can be kept with others of the same species or even close relatives, providing there is sufficient food.

HOW compatible with other invertebrates?
May be unduly bothered by large hermit crabs or other sizable crustaceans.

WILL fish pose a threat?
Occasionally fish will peck at the feeding arms. Whether this is through aggression, inquisitiveness or simply to remove captured food from the feeding arms is not clear.

WILL it threaten fish?
If stressed it can release large amounts of the sea cucumber defence toxin holothurin, that has the potential to poison and kill fish.

WHAT to watch out for?
Healthy specimens are often recent imports and well acclimatised. If their arms are extended and appear to be feeding, they are safe to buy. The longer they remain in the dealers tank without receiving suitable foods, the less likely they are to thrive. Although these cucumbers crawl slowly, they are frequently sucked into pump intakes and aquarium outlets.

WILL it reproduce in an aquarium?
Individuals sometimes release eggs. These are said to be toxic to fish, which occasionally consume them.

SIMILAR SPECIES
The royal, or purple, sea apple (*Pseudocolochirus* spp.) is a magnificent creature and demands a high price. It is just as likely to cause problems when stocked into a reef aquarium.

Synapta spp.

Feather sea cucumber

PROFILE

These bizarre animals, also known as Medusa worms, are very different from all their close relatives sold for the aquarium trade. Their bodies are like a sticky, flaccid bag, capable of stretching to a sometimes alarming degree. They are fantastic to observe, but house them in a specialised aquarium, as they often fall victim to the various items of hardware required to provide optimum conditions in a reef system.

WHAT size?
Large individuals may reach 2m when fully extended.

WHAT does it eat?
This echinoderm feeds primarily on small particulate items, such as detritus, 'marine snow' and plankton. If the aquarium is large enough, it may never require feeding specifically, but offering commercial invertebrate diets is useful.

WHERE is it from?
Tropical Indo-Pacific.

WHAT does it cost?
★★☆☆☆
Relatively inexpensive.

▶ *An active feather cucumber is one of the most bizarre sights in the animal kingdom. Provide it with a purpose-designed aquarium.*

HOW many in one tank?
Can be kept with others of the same species or even close relatives providing there is sufficient food.

HOW compatible with other invertebrates?
Capable of knocking over unattached invertebrates.

WILL fish pose a threat?
Some fish may occasionally nip at this sea cucumber. Identifying particular species is difficult, as almost any fish may be guilty at one time or another. It is likely to depend on the character of the individual concerned.

WILL it threaten fish?
Yes. When stressed, it has the potential to release toxins that can be lethal to fish. Responsible for many rapid fish die-off events in the reef aquarium.

WHAT to watch out for?
Specimens are very active and will do many laps of the aquarium as they feed. They soon find unprotected circulation pumps and filter inlets, which can suck them inside and damage or kill them. This can result in the release of their poisonous chemicals.

WILL it reproduce in an aquarium?
Unlikely, but could not be ruled out where conditions allow. Smaller species are known to reproduce asexually.

SIMILAR SPECIES
Small species such as *Synaptula* are sometimes introduced with certain corals or sponges. All are members of the sea cucumber Order Apodida and all have the same basic body plan and aquarium requirements.

Special cases

This section examines those animals that are sufficiently few in number, at least in their availability to aquarists, to justify a section to themselves or that are significantly different from other members of their grouping in terms of their aquarium demands. Tunicates, often known as sea squirts, are relatively infrequent imports for the aquarium trade, despite being fascinating creatures in their own right. Although they superficially resemble sponges they are much more complex in design. Also included here are more members of the Cnidaria, namely some hydrozoans and jellies, despite their close relationship to anemones, stony corals, soft corals and polyps. We also reserve a place for a mollusc, the Caribbean octopus. Its husbandry is so far removed from that of other molluscs in a reef aquarium that it is given special treatment here.

Price guide

★	£7 – 10
★★	£10 – 20
★★★	£20 – 25
★★★★	£25 – 35
★★★★★	£35 – 40

PROFILE

A small and infrequent import from the Philippines and surrounding region. This tunicate is one of the few species available to hobbyists that contains symbiotic photosynthetic bacteria, which account for the animal's attractive green hue. Specimens offered for sale are often very small (less than 10mm in diameter) and sometimes attached to non-secure substrates, such as macroalgae. If their living foundation dies in the aquarium, they too are unlikely to survive unless they relocate themselves before being subjected to the vagaries of the aquarium's currents.

WHAT size?
Usually grows no taller than 2.5cm.

WHAT does it eat?
Its symbionts provide it with a certain amount of nutrition, but also offer fine particulate microplankton substitutes on a daily basis.

WHERE is it from?
Tropical Indo-Pacific.

WHAT does it cost?
★☆☆☆☆ ★★☆☆☆
Depends on the number of tunicates and the size of the substrate they are attached to.

Atrolium robustum

Green urn tunicate

HOW compatible with other invertebrates?

May cause irritation to its neighbours, especially when relocating to an alternative position in the aquarium. At this time, it may also fall victim to stinging neighbours.

WHAT water flow rate?

Indirect flow is best for most individuals.

HOW much light?

Moderate to strong lighting is important for this animal, preferably T5 or metal-halide.

WILL fish pose a threat?

Avoid keeping this tunicate with angelfish, particularly if the tunicates are very small. Possible exceptions are the zooplanktivorous angelfish of the genus *Genicanthus*.

SIMILAR SPECIES

A similar species called *Didemnum molle* is very occasionally available and grows slightly larger. Both species may be encountered as accidental arrivals on coral base rock.

WILL it threaten fish?

No.

WHAT to watch out for?

The main problem with the green urn tunicate is the chance of overlooking it in a dealer's tank due to its small size. Specimens may be offered for sale on pieces of substrate only 6-7cm across. Look upon them as long-term additions that provide a stunning display when they reach their maximum size.

WILL it reproduce in an aquarium?

Where animals have plenty of food and enjoy excellent water quality, they have the potential to reproduce asexually. Daughter animals often drop away from the main colony and, hopefully, settle and colonise other areas.

▼ *These tiny yet beautiful invertebrates are currently little known in the hobby, but can prove extremely rewarding to keep.*

Clavelina robusta

Crystal tunicate

One of the few species of colonial tunicate available in the aquarium trade, albeit sporadically. Many do not survive the rigours of collection and shipping. However, if their specific demands are met in the home aquarium, they can prove hardy and long-lived. The rings around the siphons can be yellow, green or white.

WHAT size?
Each individual animal achieves around 7-8cm in height and about 10mm in diameter. Colonies can number several hundred individuals, but most of those imported for the aquarium trade are much smaller.

WHAT does it eat?
Consumes fine particulate material, including phytoplankton and suspended bacteria. To meet the animal's long-term nutritional requirements, provide slightly larger-particle foods; oyster eggs, rotifers and similar formula diets are more suitable.

WHERE is it from?
Tropical Indo-Pacific.

WHAT does it cost?
★★★★☆
Price depends on the source and size of rock.

HOW compatible with other invertebrates?
As with other encrusting organisms that reproduce asexually, good growth in healthy colonies can mean competition for substrate. But it presents little direct threat to its neighbours.

▲ *This form of clavelina is frequently offered for sale.*

WHAT water flow rate?
Indirect flow is best for most individuals.

HOW much light?
Appears to prefer lower-light situations, but can thrive in an aquarium with high wattages of metal-halide lighting, providing it is kept free of fouling algae.

WILL fish pose a threat?
Vulnerable to angelfish of any species. May be nibbled by any browsing species, so best kept with open water-swimming species that feed on zooplankton.

WILL it threaten fish?
No.

WHAT to watch out for?
Acclimatise specimens slowly and carefully. Do not expose them to the air, as this can become trapped inside and result in infections that can kill the animal.

Avoid colonies where some individuals are dying off. Timing a purchase can be difficult, as they should be fully recovered from transportation, but not allowed to starve in the dealer's tank. Good retailers will, of course, provide suitable foods where possible.

WILL it reproduce in an aquarium?
Where animals have plenty of food and enjoy excellent water quality, there is the potential for them to breed sexually and for the young to survive.

SIMILAR SPECIES
This tunicate is similar in form to many other species from the genus *Clavelina*, generally experienced as accidental imports on the base of corals and in association with live rock.

Nephtheis fascicularis

Blue palm, or blue lollipop, tunicate

PROFILE

This species is one of the most beautiful of all tunicates sold for the home aquarium, but finding it will depend on luck and where you live. It is more commonly imported into North America than Europe, but is available to the persistent aquarist. It is readily distinguished from most other colonial tunicates by the long stalks that are home to each cluster of zooids.

WHAT size?
In the wild it can form large spreading colonies over several square metres, but this is unlikely to occur in the aquarium, where it usually grows to 12cm.

WHAT does it eat?
Fine particulate material, including phytoplankton, although the nutritional value of the latter is not sufficient on its own. Slightly larger-particle foods, such as oyster eggs, rotifers and similar formula diets are more suitable. Stirring up sand beds can release bacteria and other super-fine particle foods.

WHERE is it from?
Tropical Western Pacific.

WHAT does it cost?
★★★★☆ ★★★★★
Commands a high price when available.

HOW compatible with other invertebrates?
Should not present any problems to other invertebrates unless they are immediately downstream or in physical contact. Most tunicates secrete chemicals that prevent them becoming fouled by encrusting organisms, such as algae, bryozoans and similar creatures.

WHAT water flow rate?
Indirect flow is best for most individuals.

HOW much light?
Can be maintained in high light conditions.

WILL fish pose a threat?
May be attacked by any browsing fish species, particularly angelfish.

SIMILAR SPECIES
Few species commonly available in the trade will be confused with this animal. Other members of the genus *Nephtheis* are very occasional importations.

WILL it threaten fish?
No.

WHAT to watch out for?
Specimens can shed heads in the aquarium, which is said by some authorities to be a natural process. Indeed, it may be a method of colonising new areas. On the other hand, it might indicate a stressed or ailing individual. Shedding frequency differs but can be as often as every couple of weeks.

WILL it reproduce in an aquarium?
Has the potential to proliferate but reports are sketchy, probably due to this animal's poor survival rate in captivity.

◀ *The 'blue' in the common name of this animal refers to the colour of the stalks on which the golf ball-like zooid colonies are located.*

Polycarpa aurata
Purple line tunicate

PROFILE

A large and beautiful species of sea squirt. It is probably the most frequently encountered non-accidental import from this group of animals, but often poorly understood by those who attempt to maintain it in the home aquarium. However, given the relative abundance of specialist diets for filter-feeding marine organisms available in the trade, species such as this should become much easier to maintain in the home aquarium.

WHAT size?
Usually grows no taller than 10cm in the aquarium, but has been reported achieving a diameter of 20cm.

WHAT does it eat?
Consumes fine particulate material, including phytoplankton. To meet the animal's long-term nutritional requirements, provide slightly larger-particle foods; oyster eggs, rotifers and similar formula diets are more suitable.

WHERE is it from?
Tropical Indo-Pacific.

WHAT does it cost?
★★☆☆☆ ★★★★☆
Price depends on the number on the substrate. The price range here will usually buy rocks with 1-5 individuals present.

HOW compatible with other invertebrates?
Capable of secreting chemicals that inhibit the growth of other animals – particularly those that might want to colonise the animal itself. This may retard the growth of close neighbours or animals placed downstream.

WHAT water flow rate?
Indirect flow is best for most individuals.

HOW much light?
Lighting that is too strong can cause algal growth to colonise and choke the rock on which the tunicate is imported. Introducing

▼ *Avoid specimens that appear to be secreting mucus or are deflated. This may be temporary or signify a problem.*

suitable herbivorous animals and controlling excess nutrients are important ways of preventing this.

WILL fish pose a threat?
Dwarf angelfish may nip at the flesh of this animal, as too can almost any species that browses from rocky substrates. True angelfish are known to include tunicates in their natural diets.

WILL it threaten fish?
No.

WHAT to watch out for?
Specimens should appear well inflated with vibrant coloration. Ignore those that appear excessively wrinkly or flaccid. They will often slough a mucus layer when stressed, so if this is present leave them alone to acclimatise and recover.

WILL it reproduce in an aquarium?
Where animals have plenty of food and enjoy excellent water quality, there is the potential for them to breed sexually and for the young to survive.

Rhopalaea spp.
Royal blue tunicate

PROFILE

This striking tunicate was formerly known as *Clavelina caerulea*. It is highly sought-after for the aquarium, even though reports suggest inconsistent success in maintaining it long term. In common with many tropical marine animals offered for sale that require regular feeding with plankton substitutes, there is still some way to go before most aquarists can keep this species successfully.

WHAT size?
Each animal has the potential to reach to 4cm in diameter. Specimens offered for sale are usually much smaller. They form colonies reaching about 40cm in diameter.

WHAT does it eat?
Provide a variety of proprietary zooplankton substitutes, supplemented with regular additions of brineshrimp nauplii, cyclops and rotifers where possible. Liquidised shellfish and whole fish may also be useful.

WHERE is it from?
Tropical Indo-Pacific; usually sourced through the Philippines.

WHAT does it cost?
★★☆☆☆   ★★★★★
Price depends on the size of the colony.

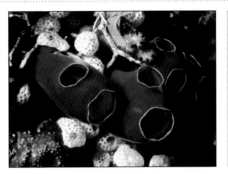

▲ *The stunning colours of this animal endear it to aquarists, but it is not straightforward to keep.*

HOW compatible with other invertebrates?
May cause irritation to its neighbours, especially when relocating to an alternative position in the aquarium. At this time, it may also fall victim to stinging neighbours.

WHAT water flow rate?
Indirect flow of moderate intensity.

HOW much light?
Does not enjoy strong illumination. Keep in shaded areas of the aquarium.

SIMILAR SPECIES

Resembles species from the genus *Clavelina,* but usually easy to distinguish by its vivid blue coloration. Its alternative common name is cobalt sea squirt.

VITAL LINK

The tadpole larvae of tunicates have been used to make a link between these animals and all vertebrates, including human beings!

WILL fish pose a threat?
This animal contains strong defensive toxins. It is unlikely that any fish will nip at it.

WILL it threaten fish?
Should not harm fish.

WHAT to watch out for?
Animals can be slightly variable in coloration, even within the same colony; this is entirely natural. Even transparent individuals may be present. Avoid newly imported specimens that have not had time to settle or those with a large number of flaccid animals present in the colony. Only purchase specimens attached to a piece of hard substrate, as loose specimens are very difficult to site and care for.

WILL it reproduce in an aquarium?
Healthy, well-fed specimens may reproduce sexually. Female individuals brood eggs internally and release well-developed larvae that are ready to settle and begin a new colony.

Cassiopea andromeda

Upside-down jelly

PROFILE

Cassiopea is the only readily available jelly in the aquarium trade. This marvellous creature contains symbiotic zooxanthellae in its tentacles, in addition to the stinging cells for which most of its relatives are famous. In order to maximise the photosynthetic efficiency of these algae, the animal spends much of its life in an inverted position, with its tentacles facing upwards, hence its common name.

WHAT size?
Will reach 30cm in diameter. Most specimens offered for sale are much smaller, measuring 10cm or so.

WHAT does it eat?
Although it is primarily able to gain nourishment via its symbionts, this jelly will also accept fine particulate foods. It ensnares these with its tentacles and passes them to its central mouth.

WHERE is it from?
Tropical Indo-Pacific.

WHAT does it cost?
★★☆☆☆
Price depends on size.

▶ *The cassiopea jelly at rest upside-down on the substrate. The photosynthetic algae borne by the tentacles are exposed to the light.*

HOW compatible with other invertebrates?
Can sting its neighbours or practically any tankmates when it moves around the aquarium.

WILL fish pose a threat?
Fish may attempt to nibble this animal's tentacles when first introduced to the aquarium, but this is quite rare. Most will avoid it or at worst attempt to remove food from the tentacles.

WILL it threaten fish?
Yes. The stinging cells are capable of causing severe injuries to fish, should they come into contact with them. Most are sufficiently wary to avoid them.

WHAT to watch out for?
Avoid direct contact with this animal or use nets or gloves when moving it. Cassiopea does not

SIMILAR SPECIES
Few species of jelly make their way into the hobby. Those that do generally need specialist care if they are to survive for any length of time.

like strong currents. Ideally, place it in its own, strongly illuminated aquarium, with others of its own kind. Cannot be stocked into systems with unprotected weirs or pump intakes.

WILL it reproduce in an aquarium?
Unlikely. In common with many jellies, this species has a hydroid stage that is sessile and capable of reproducing asexually. However, the medusoid (jelly) stage breeds sexually.

Cavernularia spp.

Sea pen

PROFILE

Sea pens are specialised corals with particular demands in an aquarium environment. Although commonly offered for sale, they are not easy to maintain at present. Given the rapid advances in coral nutrition and the increased availability of zooplankton substitutes in the hobby, it may become possible to keep them more successfully in the future. Meanwhile, there is a strong argument for not importing these animals in the first place.

WHAT size?
Including the fleshy 'foot' that remains in the substrate, this animal can grow to over 45cm. Much of this is hidden in soft substrates; the polyp-bearing crown only measures around 25-30cm at most.

WHAT does it eat?
Does not contain photosynthetic symbionts. Requires regular feeding with a variety of proprietary zooplankton substitutes. It may take food as large as frozen marine copepods or brineshrimp larvae.

WHERE is it from?
Tropical Indo-Pacific.

WHAT does it cost?
★★☆☆☆   ★★★★☆

HOW compatible with other invertebrates?
If the sea pen is secure it should not harm its tankmates. However, if it is unable to anchor itself in the substrate, it can roll around in the currents and potentially damage, or be damaged by, its neighbours.

WILL fish pose a threat?
Fish may remove captured food from the sea pen's polyps. Some angelfish may occasionally nip at their tentacles.

WILL it threaten fish?
No. This animal does not harm any fish in the aquarium.

WHAT to watch out for?
Cycles of activity are common.

◀ Sea pens are beautiful creatures, but they require such specific conditions that they are not suitable for the majority of reef aquariums.

Sea pens may not extend their polyps all day. They must have substrate to burrow into. A 20-30cm-deep layer is sufficient, but this is a great deal more than is normally found in most reef systems. Sea pens can be located in tubs or pipes filled with this depth of sand, but finding the correct location for these is vital if they are to stay put. Sea pens do not enjoy strong illumination, but require consistent, indirect currents.

WILL it reproduce in an aquarium?
Spawning in the aquarium is very unlikely given the lack of specimens thriving at present.

SIMILAR SPECIES
Few other species of sea pen are offered with any regularity in the hobby. *Pteroeides* spp. look like feathery plumes when they emerge from the sand and have small polyps. They are as demanding as all other sea pens.

Limulus polyphemus

Horseshoe crab

Although commonly referred to as a species of crab, this arthropod is a member of the Class Chelicerata and more closely related to spiders and scorpions. It has survived essentially unchanged for millions of years and makes an unusual aquarium animal, providing its requirements can be met.

WHAT size?
This is a large animal that achieves an impressive 60cm total length.

WHAT does it eat?
The horseshoe crab is a scavenger and detritivore. It will consume meaty foods that should be offered at a size suitable for the age of the animal and placed on or in the substrate. It also consumes interstitial and burrowing animals in sand beds.

WHERE is it from?
Tropical Atlantic.

WHAT does it cost?
★☆☆☆☆   ★★☆☆☆
Price depends on size.

▲ *The horseshoe crab grows large and yet can be purchased when very small. Buyer beware!*

HOW compatible with other invertebrates?
Consumes many beneficial animals in the substrate, particularly those naturally occurring species that multiply in deep sand beds. Will not harm corals directly, but is inherently clumsy in the confines of an aquarium and may undermine or overturn corals and clams through its burrowing activity.

WILL fish pose a threat?
Very few fish will attack this horseshoe crab. Those that might are seldom if ever maintained in a reef aquarium.

WILL it threaten fish?
No. This animal is entirely benign towards fish.

WHAT to watch out for?
Given sufficient food, this oddity will grow so large that it can become incredibly destructive as it digs through the substrate. It is best suited to a specialised aquarium with low flow and few obstructions to the sand bed.

WILL it reproduce in an aquarium?
No. Horseshoe crabs are famous for the mass-mating and egg-depositing aggregations that occur on the Eastern Atlantic coast of North and Central America.

SIMILAR SPECIES
There are two genera and three species of horseshoe crabs found in Asian waters. The mangrove horseshoe crab (*Carcinoscorpius rotundicauda)* grows to about 25cm in diameter. Its close relative, the coastal horseshoe crab *(Tachypleus gigas),* is larger at 30cm.

Millepora spp.

Fire coral

PROFILE

Even though they secrete a stony skeleton like many scleractinian 'true' corals, these beautiful beige-yellow animals are actually colonial hydroids. Some species contain several forms of stinging cell, some of which can inflict painful injuries if carelessly handled. Nevertheless, aquarium specimens are undervalued corals and very useful for creating systems that echo natural reefs.

WHAT size?
Can potentially grow very large – to over 90cm across – but aquarists are likely to take cuttings before then.

WHAT does it eat?
Uses its hairlike stinging cells to trap tiny plankton from the water column, but also contains symbiotic algae that provide it with the bulk of its dietary needs.

WHERE is it from?
Tropical Indo-Pacific.

WHAT does it cost?
★★☆☆☆   ★★★☆☆
Most specimens available for sale are likely to be cultured, although wild-collected individuals are sometimes available. Their higher price will reflect this.

HOW compatible with other invertebrates?
Obviously capable of stinging neighbouring corals or any animal that strays too close. Allow plenty of room for growth.

WHAT water flow rate?
Moderate to strong, indirect flow.

HOW much light?
Prefers strong illumination. In the wild, the coral is frequently found in shallow water on reefs and the lighting should reflect this in the aquarium.

WILL fish pose a threat?
Its stinging defences should deter many aquarium fish, but most reef-compatible species would not bother it anyway.

SIMILAR SPECIES
There are a few species of *Millepora*, but only one is usually offered for sale and then only sporadically. There are branching and encrusting forms. It could be confused with almost any branching small-polyp stony coral.

WILL it threaten fish?
Can potentially sting fish that stray too close, but most have an innate awareness of the powers of this coral and keep a respectful distance.

WHAT to watch out for?
Avoid specimens with bare patches of skeleton. However, white tips are normal; these are the areas where new skeleton growth occurs.

WILL it reproduce in an aquarium?
Reproduction is rare, as few aquarists keep this animal in their aquarium.

◁ *It may not be a true coral, but the fire coral's aquarium characteristics and requirements are practically identical to those of many small-polyp stony corals.*

Freeloaders in the aquarium

When placing any piece of coral or live rock into the aquarium, you are likely to be introducing not just one specimen, but potentially hundreds more living on or in it. Here, we highlight some of the more common incidental species that hitch-hike into the aquarium in this way and examine the possible consequences of their arrival. Current thinking in reef aquatics suggests that the more diverse the life in the aquarium, the more successful it will be. Many accidental arrivals into the home aquarium are highly useful, grazing nuisance algae or scavenging uneaten food intended for fish. Some may be quite benign without ever contributing to the well-being of the aquarium. Inspecting the life associated with a specimen coral can provide valuable clues to its position in its natural reef environment and thus help you decide where to place it in the aquarium.

Price guide

★	Up to £10
★★	£15 – 20
★★★	£21 – 40
★★★★	£41 – 80
★★★★★	£80+

PROFILE

Aiptasia, also known as the triffid anemone, is arguably the most commonly encountered and troublesome of all accidentally imported animals familiar to reef aquarists. It is ideally suited to reef aquariums, because they are brightly lit and because food is regularly added to them.

WHAT size?

Individual animals can grow to around 10cm in height and around 8cm across the tentacles. They may be present in the aquarium in a variety of sizes. Their ability to reproduce by asexual means through pedal laceration and fission result in the formation of huge colonies that can dominate an aquarium. (Pedal laceration is the voluntary 'tearing' of the foot with the resultant tissue fragments being able to develop into new animals.)

WHAT does it eat?

Small, meaty particulate material. Enjoys almost any kind of feed intended for fish or invertebrates. Also contains zooxanthellae.

WHERE is it from?

Circumtropical.

Aiptasia, or glass rose, anemone

▲ *Aiptasia are amongst the most common and troublesome accidental imports into the marine aquarium.*

WHAT potential harm can it cause?

Apart from being unsightly, this anemone has a potent sting. The fact that it reproduces prolifically means that neighbouring sessile invertebrates can be severely damaged if and when they come into contact with it and are stung. It will often out-compete most other sessile invertebrates for substrate.

WILL fish pose a threat?

Some butterflyfish, including *Chaetodon burgessi, C. striatus* and *Chelmon rostratus,* will nip at the anemone's tentacles, causing it to contract. However, these fish may also predate other animals that the aquarist wishes to maintain, such as tubeworms.

WILL it threaten fish?

It can inflict painful wounds on fish. It may even catch and eat small fish such as gobies.

WHAT control methods?

An increasing number of off-the-shelf treatments are available for tackling this pest anemone and most work well, if only in the short term. Peppermint shrimp *(Lysmata wurdemanni)* are known to consume aiptasia. The ultimate controller is thought to be the sea slug *Aeolidella stephaniae* (formerly known in the aquarium trade as *Berghia verrucornis*). This nudibranch mollusc has been bred and successfully reared in captivity. It only eats aiptasia. Quarantining new corals can enable aquarists to locate new specimens of aiptasia before they are introduced to the aquarium.

WILL it reproduce in an aquarium?

Reproduces asexually.

UNDESIRABLE ADDITIONS

Of course, other, more sinister elements may arrive too, and they should be removed where possible. However, unless you can establish direct harm to another aquarium resident, do not remove or destroy unidentified animals indiscriminately. Where possible, it is a good idea to move them to a separate aquarium for observation.

▲ *Anemonia cf.* majano *is another pest anemone species that thrives in the high light, abundant food environment of the reef aquarium.*

SIMILAR SPECIES

Another problem anemone is the majano, or manjano *(Anemonia* cf. *majano),* an animal that is reminiscent of very small specimens of the bubble anemone *(Entacmaea quadricolor).* This species usually remains fairly small – perhaps a few centimetres tall and 20mm or so across the tentacle span. However, it has the potential to reproduce almost as quickly as aiptasia. The lace anemone *(Thalassianthus aster)* is another potentially problematic anemone, but is distinct due to its highly branched tentacles. It appears to be less hardy than the former species and easier to control using proprietary treatments.

Asterina spp.

Asterina starfish

PROFILE

A genus of small starfish that look as though they have been half-chewed by some invisible predator. They can have varying numbers of legs – from three to seven or more – some of which will be much shorter than the others. At least three species are known, but they are often grouped together when being described. This would be acceptable were it not that one species appears to predate certain corals, whereas the other two are apparently harmless.

WHAT size?

Reaches around 2cm across the arms when fully grown.

WHAT does it eat?

Green-backed specimens of asterina have been observed attacking a variety of stony corals, including *Acropora* spp. Asterina with beige or predominantly cream coloration have been observed consuming dying or ailing polyps, but appear to prefer scavenging uneaten food intended for fish or consuming algal films on rock or glass.

WHERE is it from?

Tropical Indo-Pacific.

▶ *The arms on this specimen (seen from below) are roughly equal in length but in many individuals they are more irregular.*

WHAT potential harm can it cause?

The species or individuals that appear to predate corals strip away patches of tissue, leaving exposed white skeletal material. Establishing the guilt or innocence of a specimen of asterina should not be difficult, given a little patience and careful observation. Without the aquarist's intervention, a single starfish can strip an entire colony.

WILL fish pose a threat?

Although fish that eat this starfish are almost certain to exist, there are no reliable reports of them being predated in the home aquarium.

WILL it threaten fish?

No.

WHAT to watch out for?

Most asterina starfish are completely harmless and should

be welcomed into the reef aquarium for the beneficial role they undertake. The proper maintenance of any aquarium involves observing the inhabitants closely and checking for signs of problems or malaise. This should quickly reveal any problems with starfish. Remove them with tweezers or carefully by hand.

WILL it reproduce in an aquarium?

Yes. Asexual reproduction is common and can continue apace. Some aquarists have been forced to resort to the starfish-eating shrimp *Hymenocerca elegans* in order to control the numbers of starfish.

Convolutriloba retrogemma

Rust flatworm

PROFILE

A common species of pest organism that is introduced with live rock and/or coral base rock. Its small size belies the fact that it can form large colonies and overwhelm a reef aquarium. The flatworms are a distinctive shape, usually with an obvious orange-coloured central area of the body.

WHAT size?

Individuals usually measure less than 5mm. Given suitable conditions, huge colonies can form, numbering several thousand individuals. These will smother almost any available substrate, including some corals, as well as sand and base rock.

WHAT does it eat?

Ingests the zooxanthellae from damaged corals, particularly *Discosoma* and *Actinodiscus* mushroom polyps. These are assimilated into the tissues, where the photosynthetic algae use the light incident upon them to nourish the flatworm. The latter can be observed stretching from the substrate to offer the maximum surface area for light absorption.

WHERE is it from?

Tropical Indo-Pacific.

▲ *Population explosions in this species are difficult to anticipate or halt once in progress. It is hard to sit back and watch, but this often proves the best course, providing there is no direct threat to aquarium inhabitants.*

WHAT potential harm can it cause?

Has the potential to smother corals, thus depriving them of light, or at least irritate them when they are in close proximity. Otherwise, the flatworm is very unsightly.

WILL fish pose a threat?

Some fish will consume the flatworm, including members of the genus *Synchiropus* (dragonets such as the spotted mandarin, *S. picturatus*). However, this often depends on the individual fish concerned. Do not stock it for control reasons alone, as many will completely ignore the flatworm.

WILL it threaten fish?

Rapid die-off amongst the flatworms can result in the release of large quantities of

toxins into the water. These can directly harm both fish and invertebrates and also reduce oxygen levels as the flatworms decompose. For this reason, few people are willing to try any of the limited number of off-the-shelf treatments available for this problem animal.

WHAT control methods?

Patience is the key with this animal. Siphoning off individuals during water changes (using a 6mm airline) helps keep their numbers in check, particularly if they are beginning to smother corals. However, doing nothing will often have the same result. Over a matter of weeks to months, the population of flatworms will usually diminish. Quarantining corals can be useful in identifying rocks that are home to this pest

WILL it reproduce in an aquarium?

Reproduces asexually through bilateral fission.

SIMILAR SPECIES

Another flatworm *(Waminoa sp.)* is usually brown, with a shape reminiscent of a cross-section through an apple. It can infest corals with varying degrees of severity. Freshwater dips are recommended for particularly severely affected coral specimens.

Echinometra spp.

Black sea urchin

A small species of sea urchin that is initially welcomed as a 'freebie' when first noticed by the aquarist. However, in many cases it soon outlives its welcome, due in part to a prodigious growth rate. It is usually identified through the white rings at the base of the spines and the spherical body.

WHAT size?
10-12cm spine diameter.

WHAT does it eat?
Primarily algae. Eagerly consumes film, filamentous and calcareous forms. Can be offered dried forms or pellets and it may consume uneaten food intended for fish.

WHERE is it from?
Tropical Indo-Pacific.

WHAT does it cost?
☆☆☆☆☆
Unlikely to be offered for sale. Usually free.

HOW compatible with other invertebrates?
Not likely to consume or damage any aquarium inhabitants intentionally, but is clumsy and surprisingly strong. It can overturn animals on its nocturnal forays.

WILL fish pose a threat?
Few fish likely to be kept with live rock will tackle a sea urchin. A possible exception may be the triggerfish if kept in a live rock-based system. It 'blows' jets of water at the base of the urchin in an attempt to overturn it and expose its less well-defended underside.

WILL it threaten fish?
No.

WHAT to watch out for?
This species arrives when very small but grows quickly. It has the potential to be a useful herbivore but often leaves unsightly white

The very similar *Parasalenia gratiosa* feeds on stony coral polyps, but does not grow as large as *Echinometra* (3-4cm across the spines) and appears to have a more flattened body. It is best to remove any small, accidentally acquired urchins to prevent potential problems later on.

patches where it has removed colourful calcareous algae from the rockwork. Offering cuttlebone for it to rasp can deter this behaviour.

WILL it reproduce in an aquarium?
Sexes are separate. May spawn, sometimes even synchronously, but any resultant larvae will not survive.

▶ *This attractive urchin bores holes in rockwork in its natural environment.*

Euplica varians, Euplica spp.

Dove snail

PROFILE

Dove snails are highly desirable hitch-hikers that are extremely useful herbivores in the home reef aquarium. They are occasionally offered for sale, but arrive by accident on specimen pieces of coral or live rock.

WHAT size?
The shell of this small snail measures 10-12mm at most.

WHAT does it eat?
Microalgae form the majority of the diet of this gastropod mollusc. It will consume diatoms, but also green and brown forms that grow on glass and rockwork.

WHERE is it from?
Tropical Indo-Pacific.

WHAT does it cost?
Usually free, but can occasionally be bought very cheaply.

SIMILAR SPECIES
There are several species of gastropod mollusc with a similar shell shape. Many such animals are benign, but it is possible that some carnivorous whelks might be mistaken for the beneficial dove snail.

HOW compatible with other invertebrates?
Dove snails present no threat to other invertebrates, but where established populations occur they may compete for food with other herbivorous snails, such as *Turbo, Trochus* and *Astrea.*

WILL fish pose a threat?
Small wrasse, such as members of the genus *Pseudocheilinus* and *Halichoeres,* are known to predate small snails. Many other fish are likely to consume small juveniles as they emerge from their egg capsules.

WILL it threaten fish?
This snail presents no threat to any fish species.

WHAT to watch out for?
These wonderful snails will reproduce prolifically in the aquarium. Their numbers and

▲ *The dove snail is small enough to be overlooked by aquarists, but can form large populations and consume significant quantities of algae if left unmolested.*

small size can mean that they get into small apertures and can block pumps or outlets if steps are not taken to prevent this. However, as a species their benefits greatly outweigh these disadvantages.

WILL it reproduce in an aquarium?
Yes. Following mating, a small number of eggs are deposited inside a transparent capsule, often visible on the aquarium glass. The larvae undergo direct development, which means that there is no planktonic life-cycle stage and they emerge as miniature copies of the adults.

Eurythoe spp.

Polychaete bristleworms

PROFILE

Historically, bristleworms have received a very poor press, based on a number of incidents with directly harmful species or with large and hungry members of usually more benign species. Today, most aquarists seem to recognise these accidental yet inevitable imports for the benign or potentially beneficial animals they are.

WHAT size?

Can achieve 45cm in length. Most species found in aquariums are smaller, with more modest appetites. Some live in sand or other soft substrates, whereas others prefer holes in rockwork.

WHAT does it eat?

Primarily scavengers and detritivores. Will consume algae and uneaten food intended for fish and invertebrates. Occasionally takes advantage of ailing or dying fish or corals.

WHERE is it from?

Circumtropical.

▶ *Free-living 'bristleworms' are generally centipede-like in appearance and may be found inhabiting rock or sand/gravel substrates. Several species are known be introduced unintentionally, the majority of which are benign.*

WHAT potential harm can it cause?

Large and hungry individuals will occasionally take bites out of sessile invertebrates and should be removed. Small bristleworms can be predated by the arrow crab (*Stenorhynchus seticornis*).

WILL fish pose a threat?

Several fish species are known to consume bristleworms, including some members of the genus *Pseudochromis* (the dottybacks) and several butterflyfish.

WILL it threaten fish?

Larger specimens can catch and eat fish, but these are usually ailing in the first place to allow themselves to be caught.

WHAT to watch out for?

In many species, hairlike projections called setae arise from the body segments, hence the common name 'bristleworm'. Often, these contain a venom

SIMILAR SPECIES

Two predatory species are worth mentioning. *Oenene fulgida* is a bright pink-orange worm known to predate tridacnid clams and gastropod molluscs, such as *Turbo* spp. or *Astrea* spp. It differs from the beneficial similarly coloured species by having barely noticeable setae. *Hermodice carunculata* is a beautiful-looking worm with prominent and ornate gills along its flanks. It consumes a wide variety of corals. Remove both species carefully.

capable of causing severe pain or irritation when it comes into contact with an aquarist's skin, hence the alternative common name of fireworm. Many reef-keepers use gloves when handling rocks and corals. Handling the worms carelessly can cause them to break into two or more pieces, which can then regenerate into new animals.

WILL it reproduce in an aquarium?

Yes. Can reproduce asexually. Usually, a stable population forms according to the amount of food available in the aquarium. Occasionally, modified sexual individuals called epitokes appear. These are often fast-swimming and dart around the aquarium before dying as they release sperm or eggs.

Heliacus spp.

Chequered sundial snail

PROFILE

This species of marine gastropod predates zoanthid button polyps and should be removed whenever encountered. Fortunately, it is one of the least subtle animals that aquarists are likely to encounter as an accidental import, as it has a tendency to be located in the centre of a patch of its favourite food. This characteristic, plus its attractive striped coloration, mean that it is seldom difficult to identify.

WHAT size?
Each snail reaches about 2cm across the shell.

WHAT does it eat?
Primarily zoanthids, but some species will also consume members of the genus *Palythoa*. The snail makes a hole at the base of the polyp and sucks out the contents. The remaining 'husk' can offer a vital clue to the presence of this predator if it has gone unnoticed by the aquarist.

WHERE is it from?
Tropical Indo-Pacific.

◀ Box snails are not masters of disguise and seldom roam too far from their next meal.

WHAT potential harm can it cause?
If left unchecked, specimens can predate and wipe out entire colonies of zoanthids. They will not harm other animals in the aquarium, but in the absence of their only known food items they will starve anyway.

WILL fish pose a threat?
Certain wrasse, including members of the genera *Macropharyngodon* and *Pseudocheilinus,* may be useful at removing small specimens of heliacus. Larger specimens are likely to have been removed by the aquarist before the rock on which they were living was introduced into the aquarium.

WILL it threaten fish?
No.

WHAT control methods?
Before buying zoanthid button polyps, always check them for the presence of heliacus.

Sometimes the snails may be snuggled down in a group of expanded polyps and therefore obscured from view. Gently running a gloved fingertip through the colony can dislodge the snail or reveal specimens.

WILL it reproduce in an aquarium?
Has been reported laying eggs in the aquarium. This is usually done within the colony of button polyps on which the adults are feeding.

SIMILAR SPECIES
There are at least eight members of the genus *Heliacus* that predate button polyps. Of these, the chequered box snail is the most distinctive but almost any snail with a particularly flattened spire and an apparent fondness for living amongst button polyps should be viewed with suspicion.

Mithrax spp.

Mithrax crab

PROFILE

Mithrax crabs are variable in appearance and coloration, and closely related to *Mithraculus* sp., the emerald crab. They share the 'spoon-shaped' pincer ends that make a useful distinguishing characteristic between these crabs and less benign species.

WHAT size?
3-5cm across the carapace.

WHAT does it eat?
All specimens will scavenge uneaten food intended for fish or other invertebrates. Their usual natural diet is opportunistic, but will include large amounts of algae that are scraped off rockwork.

WHERE is it from?
Tropical Indo-Pacific.

WHAT does it cost?
Unlikely to be offered for sale. Usually free.

HOW compatible with other invertebrates?
Reports of these crabs attacking invertebrates are sketchy and could be due to them being mistaken for other crabs known to attack corals. However, their opportunistic nature means that they will not refuse the opportunity to take advantage of an ailing specimen.

WILL fish pose a threat?
Small or newly moulted crabs are vulnerable and may be consumed by opportunistic fish, such as *Pseudochromis* dottybacks or even dwarf angelfish *(Centropyge* spp.*)*.

WILL it threaten fish?
Large specimens will catch and eat sick or dying fish. However, the timid nature of most of these crustaceans means that they usually shun any confrontations with fish.

WHAT to watch out for?
Large specimens have

correspondingly substantial appetites and may have to be target-fed to prevent them attacking other aquarium inhabitants. If a rogue specimen materialises, it can be trapped out using any of the off-the-shelf devices for this purpose. Alternatively, use a tall, smooth-sided glass, baited with shellfish.

WILL it reproduce in an aquarium?
Spawning may be possible where a number of individuals are housed. The female carries eggs beneath her tail, which is broader than the males to facilitate this. The larvae are unlikely to survive after they hatch.

◀ *Carapace and leg colour are characters that vary considerably in* Mithrax spp. *Check the pincer shape in order to determine whether an unidentified crab is a member of this largely benign genus.*

Myronema spp.

Myronema, or myrionema

PROFILE

A diminutive colonial hydroid that contains photosynthetic pigments. It has an impressive growth rate and a potent sting. However, compared with many other cnidarians encountered in reef aquariums, this species is relatively easy to eradicate. It grows on fibrous 'runners' that encrust any vacant substrate and when this runs out, it can form twisted, ropelike strands of hydroids that drift around in open water anchored at one end.

WHAT size?

Each polyp reaches about 5-6mm in height and a similar diameter, although this may increase when they are located in strong flow. Colonies appear to have no maximum potential size, as this tends to be determined by available substrate rather than species limitations.

WHAT does it eat?

Able to provide its own nutritional requirements through its photosynthetic algae. It also appears to trap fine particulate material from the water. Although it is seldom purposely fed, this animal steals food intended for fish and invertebrates.

WHERE is it from?

Circumtropical.

Fibrous 'runners' are characteristic of this prolific animal.

WHAT potential harm can it cause?

The main problem with myronema is the fact that it contains stinging cells and may therefore harm sensitive neighbours. In reality this seldom happens and many species of sessile invertebrate, such as zoanthids and *Discosoma* spp., are tolerant when in close proximity. The impact of the presence of myronema is likely to depend on the specific mix of corals and invertebrates in the aquarium.

WILL fish pose a threat?

It is possible that some species of butterflyfish might predate this hydroid, but this is unlikely among those species that are sometimes housed in the reef aquarium.

WILL it threaten fish?

Capable of inflicting stings on fish, but there are no records of these proving harmful to fish.

WHAT control methods?

Colonies can be removed physically by peeling away strands from the rockwork en masse. Many off-the-shelf treatments for cnidarian pests such as aiptasia can also work with myronema; the latter are often more sensitive to them than the animal for which they were developed. Often the colony will simply disappear over time without having ever caused any harm.

WILL it reproduce in an aquarium?

Reproduces asexually.

SIMILAR SPECIES

Resembles the sessile stage in the life cycle of the jelly *Nausithoe*. This is a fairly common accidental import in its own right, but for this animal a definite tube can be distinguished from the circle of tentacles that protrude from it. Other hydroids commonly appear on the glass of a marine aquarium when it is maturing.

Sea fleas

PROFILE

A number of small crustaceans find their way into reef aquariums with some regularity. Their correct identification is difficult and aquarists often need to adopt a 'wait and see' approach to determine the impact of such animals on their system. However, the majority of species are either benign or beneficial, being excellent scavengers and detritivores.

WHAT size?

Most miscellaneous small crustaceans will not grow to more than 10mm. Most are smaller.

WHAT do they eat?

Microalgae, detritus, bacteria and uneaten food intended for fish or corals.

WHERE are they from?

Circumtropical.

▲ Munnid isopods measure only a few millimetres in length and are commonly observed on the aquarium glass during the early life of the aquarium. Stable populations may develop in sumps and refuges.

HOW compatible with other invertebrates?

As with many invertebrate groups, there are good and bad representatives. Most aquarium isopods are harmless scavengers and detritivores, but some species parasitise other crustaceans. Some amphipods are known to attack stony corals. Fortunately, such undesirable animals are often obvious and infrequently encountered.

WHAT potential harm can they cause?

Certain isopods parasitise fish. These blood-sucking animals are very obvious when infecting their host and they usually drop off after a few days, but not until they have caused the aquarist considerable worry!

WILL fish pose a threat?

A wide range of fish consume these animals. Certain dragonets (*Synchiropus* spp.) actually depend on the incidental crustaceans in a reef aquarium for the majority of their food. Stocking several such fish

▲ This 7mm-long tanaidacean was removed from a burrow it had made in the sand of a reef aquarium. The species had formed a stable, self-perpetuating population there and it was a highly useful scavenger and detritivore.

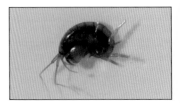

▲ Amphipods are among the most abundant of all miniature crustaceans in marine aquariums. Found on rock and sand, they are most frequently encountered on mechanical filtration media, where they consume detritus.

can lead to the collapse of an invertebrate population, with the result that the fish starve. Aquarists should remember the beneficial role undertaken by these small crustaceans when considering which species of fish to stock in their aquarium.

WHAT to watch out for?

Unless they are obviously causing harm, try to observe any accidental arrivals closely before taking action. It is also worthwhile considering the context in which the animal is encountered in order to gain clues as to their aquarium disposition. For example, certain amphipods are commonly found living on mechanical filters, where they consume detritus.

WILL they reproduce in an aquarium?

Some species can form large populations. Where conditions suit them, isopods, mysids, amphipods and tanaidaceans can number several hundred individuals.

Tritoniopsis elegans

Soft coral sea slug

PROFILE

T. elegans is a beautiful animal that most aquarists would be happy to stock into their reef aquarium, were it not for the fact that its favourite food is soft coral! Apparently, it has the ability to remain dormant for extended periods, perhaps as an egg, only emerging when in the presence of a suitable food source. It is primarily nocturnal in habits.

WHAT size?
Reaches around 6-7cm when fully grown.

WHAT does it eat?
Commonly attacks soft corals from the genera *Sarcophyton*, *Cladiella* and *Klyxum*, but may consume other species where these are not present.

WHERE is it from?
Tropical Indo-Pacific.

SIMILAR SPECIES
Several predatory nudibranchs are known to be imported with coral base rock, often living on the same rock as their prey. *Clavularia* spp., zoanthids and *Pachyclavularia* are the corals most commonly infested.

WHAT harm can it cause?
The threat to invertebrates is solely to soft corals, which will be rasped away by the slug's radula. Often, all the aquarist will see is an accumulation of powderlike deposits at the base of the coral as it begins to shrink in size. These are the indigestible calcium carbonate skeletal elements, known as scelerites. A lone slug can decimate the soft corals in an aquarium, but they are rarely encountered singly.

WILL fish pose a threat?
There are no reports of any fish finding this nudibranch palatable.

WILL it threaten fish?
No.

WHAT control methods?
This pest rarely strays far from

▲ *This orange specimen is only one of the colour morphs of* Tritoniopsis elegans. *The white form is more commonly encountered.*

its food, so be sure to check the base of newly acquired corals before introducing them into the aquarium. Quarantining corals is a useful procedure. The slugs are extremely cryptic, especially when ensconced in a crack or crevice during the day. Prise them away from the rock with a cotton bud or similar implement.

WILL it reproduce in an aquarium?
Yes. Mating occurs in the aquarium before a small number of eggs is deposited in squiggly masses. Single individuals will deposit eggs but these are infertile.

A-Z species list featuring the main entries

Credits

The publishers would like to thank the following photographers for providing images, credited here by page number and position: B (Bottom), T (Top), C (Centre), BL (Bottom left), etc.

Clay Bryce: 42, 102, 109, 111, 124, 131, 132, 141(BL), 152, 156, 166, 168, 172, 173, 180, 185, 205

Bioquatic Photo – A.J. Nilsen, NO-4432 Hidrasund, Norway (email: bioquatic@biophoto.net. Website: www.biophoto.net): 14, 21, 22, 23, 24, 26, 28, 29, 30, 31, 32, 33(both), 34, 35, 36(both), 39, 40, 41, 43, 45, 46, 47, 48, 49, 52, 53, 58(BL), 61(R), 63(T), 64, 67, 72, 73, 74, 76, 78, 80, 81(BC), 84, 85, 88, 90, 91, 92, 93, 94, 95, 96, 99, 101, 103, 105, 106, 107, 108, 112, 116, 120, 121, 122, 123, 126, 127, 129(both), 135, 137(both), 140, 141(T), 144, 145, 146, 147, 148, 149, 150, 151, 153, 154, 155, 159, 161, 162, 164, 165, 169, 171, 174, 175, 177, 182, 183, 184, 186, 188, 189, 190, 191, 192, 193, 195(TL), 196, 197, 198, 200, 203, 204(BL)

Neil Hepworth: 16

Tristan Lougher: Intro page, 6, 17, 60(T), 79, 82, 87(both), 89(TC), 110, 128, 130, 133, 178, 187, 195(TR), 199, 202, 204(BC,TR)

Scott Michael: 15, 18, 19, 20, 54, 55, 60(B), 69, 70, 75, 77, 83, 86, 89(BR), 97(TR), 98, 100, 104, 113, 117, 125, 134, 136, 138, 139, 142, 157(both), 158, 160, 163, 167, 170, 179, 194, 201

Geoff Rogers © Interpet Publishing Ltd: Title page, 4, 5, 7, 8, 10, 11, 12(both), 13, 25, 27, 37, 44, 50, 51, 56, 58(BR), 59, 61(TL), 62, 63(R), 65, 66, 68, 71, 81(R), 97(BL), 114, 115, 118, 119, 181

Robert Sutton: 38

Iggy Tavares: 57, 143, 176

Author's acknowledgements

The author would like to thank Steve, Faye and Lisa Birchall for their continued support and assistance throughout the writing of this book. Chris Higinbotham and Paul Culcheth, Rob Hill, Andy Lister, Nicola Smart and Nicola-Jayne Fitton of the Tropical Marine Centre in Manchester assisted in obtaining specimens for aquarium study.

Publisher's acknowledgements

The publishers would like to thank the following for their help in providing facilities for photography:

Cheshire Water Life, Northwich, Cheshire.

Interfish, Wakefield, Yorkshire.

Tuan Pham.

Swallow Aquatics, East Harling, Norfolk.

Swallow Aquatics, Southfleet, Kent.

Wharf Aquatics, Pinxton, Nottinghamshire.

Publisher's note